PERSPECTIVES IN
CRYSTALLOGRAPHY

PERSPECTIVES IN
CRYSTALLOGRAPHY

John R. Helliwell
University of Manchester, United Kingdom

CRC Press
Taylor & Francis Group
Boca Raton London New York

CRC Press is an imprint of the
Taylor & Francis Group, an **informa** business

The colourful crystals depicted on the front cover relate to Chapter 6 entitled "The structural chemistry and structural biology of colouration in marine crustacea." They serve as illustrations of the themes of the book including the sustainability of life (Chapter 8). The other crystals are those of 'common salt' (sodium chloride) and relate to the historical aspects of X-ray crystal structure analysis described in Chapters 2 and 3.

The pictures of the crystals shown are with the permission of the publisher International Union of Crystallography, see http://journals.iucr.org/ and of the authors, namely:

Professor Naomi Chayen re Figure 2 from Acta Cryst. (1998). D54, 8-15 doi:10.1107/S0907444997005374 Comparative Studies of Protein Crystallization by Vapour-Diffusion and Microbatch Techniques by N. E. Chayen; the β-crustacyanin (blue) protein was crystallised by Professor Chayen and the protein had been extracted and purified from the European lobster carapace by Dr Peter Zagalsky.

Dr Madeleine Helliwell who grew the various red, and one orange, crystals re the studies of the carotenoids below reported in Acta Cryst. (2007). B63, 328- 337 doi:10.1107/S0108768106052633 Unravelling the chemical basis of the bathochromic shift in the lobster carapace; new crystal structures of unbound astaxanthin, canthaxanthin and zeaxanthin by G. Bartalucci, J. Coppin, S. Fisher, G. Hall, J. R. Helliwell, M. Helliwell and S. Liaaen-Jensen.

CRC Press
Taylor & Francis Group
6000 Broken Sound Parkway NW, Suite 300
Boca Raton, FL 33487-2742

© 2016 by Taylor & Francis Group, LLC
CRC Press is an imprint of Taylor & Francis Group, an Informa business

No claim to original U.S. Government works

Printed on acid-free paper
Version Date: 20150917

International Standard Book Number-13: 978-1-4987-3210-9 (Hardback)

Visit the Taylor & Francis Web site at
http://www.taylorandfrancis.com

and the CRC Press Web site at
http://www.crcpress.com

To my mother and father; to my wife Madeleine and our son James and daughter Katherine.

In loving memory of Nick Helliwell (1983–2011).

William Blake

(1757–1827)

(From "Auguries of Innocence")

To see a world in a grain of sand
And a Heaven in a wild flower,
Hold infinity in the palm of your hand
And eternity in an hour.

Contents

SECTION I Public understanding of crystallography

SECTION II Celebrating the centennial of X-ray crystal structure analysis

SECTION III Aspects of crystallography research

SECTION IV   Societal impacts

Foreword

The International Year of Crystallography was celebrated 2014 by lectures, symposia, articles, new books and web activities. This year was chosen because Max von Laue was the first in the field of crystallography to be awarded the Nobel Prize in Physics 1914 due to his finding in 1912 that a crystal put into an X-ray beam diffracted X-rays. This not only led to the understanding that X-rays are electromagnetic radiation, but the same year William Henry and William Lawrence Bragg realized and showed that the atomic arrangement in crystals could be deduced from the intensities of the diffracted beams. In fact the celebrations of these developments began already in 2012. Among other activities a symposium was held in Adelaide to celebrate the centennial of the groundbreaking experiments by one of the sons of the city, W.L. Bragg. The lectures were published as a special issue of *Acta Crystallographica series A*.

Crystallography has been an extremely fruitful scientific field. In the beginning it gave the first glimpses of an understanding of the atomic organization of simple salts and minerals, but gradually more challenging materials could be analyzed. Not only X-rays were used, but gradually electrons or neutrons added new possibilities to investigate crystal structures. In the 1950s the interest in structures of biological systems led to extremely important steps forward, like the structure of DNA or the first protein structure. These successes have been followed by an extreme flood of structures as complex as ribosomes or whole viruses. Many of the latest developments would not have been possible without the development and usage of synchrotrons, where the radiation initially was a nuisance to the physicists who worked on them but subsequently became an enormous benefit to crystallographers and others who needed intense X-ray beams. The field of crystallography has been instrumental for numerous fields of science where the structural knowledge has led to fundamental new levels of understanding. Therefore this discipline has been awarded at least 20 Nobel Prizes in physics, chemistry and physiology or medicine.

The author of this book, Professor John R. Helliwell, has throughout his scientific career been involved in central activities in the field, not only in determining structures and involved in education, but he has had important roles in the development of experimental possibilities at synchrotrons and also at neutron sources. He has also held central positions like chief editor of the journals published by the International Union of Crystallography as well as being the chair of the European Crystallographic Association. He has written an extensive coverage of how to make best use of synchrotrons for crystallography. The current book contains on one face of it reviews that he has published in *Crystallography Reviews*, a journal for which he is the editor. These articles cover a wide range of topics including historic accounts of crystallography, the evolution and use of synchrotron radiation for crystallography, the possibility of locating the lightest atom, hydrogen, in crystallographic experiments as well as the structural biology behind the coloration of marine crustacea. The book also contains chapters for a more general audience partly focusing on the public understanding of crystallography but also on where the field may be heading and its role in the sustainability of life. It is a book addressing great challenges and is of a broad general interest.

Anders Liljas
Biochemistry and Structural Biology
Lund University
Leksand, Sweden

Preface

The International Year of Crystallography approved by the United Nations and UNESCO took place in 2014. A key message was to build on the achievements of that year. This book aims to contribute to that process. It brings together a wide range of topics to interest both specialists and non-specialists alike. The book opens with a chapter describing my own efforts at explaining crystallography and crystal structure analysis to a wide range of audiences. I then highlight the history of crystal structure analysis. This is followed by several detailed review articles which explain a representative suite of topics in the field of crystallography, concluding this section with a short description of the future of crystal structure analysis in the next 100 years. The book concludes with a chapter describing some of our inputs, as a field, to the sustainability of life.

I am especially grateful to Anders Liljas for agreeing to write the Foreword.

I am very grateful to Hilary Rowe of CRC Press for her wise insights and comments on my book project. I am grateful to the managing editor of *Crystallography Reviews*, Huw Price, for his permission to allow my reviews in *Crystallography Reviews* to be reproduced in this book. I am also grateful to Stu Fisher, Richard Henchman and James Wilkinson for their permission to include our review article as Chapter 5 in this book. I am grateful to Stuart Eyres of Daresbury Laboratory for the 'morph photo' of me onto Dali's picture *Galatea of the Spheres* (see back cover of the book). I first saw this at an exhibition of Dali's art in Rome in a gallery at the Piazza del Popolo. (I was in Rome chairing a workshop on CCD detectors for crystallography held at the Università degli Studi di *Roma* 'La Sapienza' funded by the European community.) Dali (http://en.wikipedia.org/wiki/Galatea_of_the_Spheres), recognising that matter was made up of atoms which did not touch each other, sought to replicate this in his art at the time, with items suspended and not contacting each other. This intertwining of the abstract with the science achieved by Dali in his *Galatea of the Spheres* emphasises, for me, the importance of one's ideas and imagination and the perfecting of one's skills, mathematical and experimental, to progress one's contributions to science.

A short interview with me, briefly describing the International Year of Crystallography, some of my scientific biographical details and my role as editor of *Crystallography Reviews* is available at http://www.chemistryarena.com/06/2014/uncategorized/interview-with-john-helliwell-editor-of-crystallography-reviews.

Overall, I hope that this book of perspectives will be of interest to crystallographers at all stages of their careers from graduate students and post-docs to established academics. In addition, crystallography is enjoying unprecedented public interest arising from the International Year of Crystallography. I also hope then that this book will help sustain the interest of the public and schoolchildren in crystallography.

John R. Helliwell
Emeritus Professor of Chemistry
University of Manchester
Manchester, United Kingdom
FInstP FRSC FSoc Biol Fellow of the ACA
DSc Physics
University of York
York, United Kingdom

Author

John R. Helliwell has wide experience teaching physicists, chemists and biological scientists at undergraduate and postgraduate research levels and ran a research group at the University of Manchester, United Kingdom for over 20 years which has comprised scientists worldwide. Working at the Daresbury Laboratory, up to department head level supervising over 200 staff, greatly broadened his perspectives. In both academic and scientific civil service contexts, he has presented crystallography and research to diverse audiences. He has served the International Union of Crystallography as a representative in several global organisations, including pioneering the general topics of open access of literature linked with crystallographic research raw data. Sustainability of life, mentoring and the importance of gender equality are important elements of his work and efforts. He has chaired science advisory committees in Japan, Australia, the United States and Europe. Professor Helliwell is now emeritus professor in chemistry at the University of Manchester, United Kingdom. He has a DPhil in molecular biophysics from the University of Oxford and a Doctor of Science Physics degree from the University of York. Professor Helliwell is a recipient of the Patterson Award of the American Crystallographic Association and the Perutz Prize of the European Crystallographic Association. He is a fellow of the Institute of Physics, the Royal Society of Chemistry, the Royal Society of Biology and the American Crystallographic Association.

Section I

Public understanding of crystallography

1 Explaining 'What is crystal structure analysis?' for a general audience

One of the most challenging situations is to explain crystal structure analysis to a general audience. There are different types of such audiences one can meet, which I now take in turn as to how I approached them including the most general through to quite specialist audiences.

1.1 SCHOOL OF CHEMISTRY, UNIVERSITY OF MANCHESTER OPEN DAYS

These are principally to give the opportunity for people interested in applying to read chemistry as an undergraduate degree the chance to meet staff and see a wide range of 'chemistry in action'. Mostly, the audience would be parents, with their 17-year-old children, who would be deciding to which of the five UK universities they would apply. Figure 1.1a through d show our typical display of molecular models, crystals, computer movie demos and example research or popular magazine articles. Each item on our display would be something about which members of the display team would be both knowledgeable and indeed enthusiastic. Naturally, there would be a fair fraction of research we had undertaken.

From left to right in Figure 1.1a, a side view in Figure 1.1b (with my colleague Dr G. Habash), a close up view in Figure 1.1c (including Dr Madeleine Helliwell and myself) and a more distant view in Figure 1.1d (with Dr Madeleine Helliwell and Dr Peter Skabara):

The sugar-binding proteins concanavalin A isolated from beans (Jack and the Beanstalk beans fame) and hen egg white lysozyme

Common salt model

Calcite model (to explain the planar carbonates and the calcite crystal double image refraction effect as well as each optical image being polarized light demonstrated very readily using a piece of polaroid and a calcite crystal)

Crystals of calcite, quartz, copper and a silicon wafer

Molecular models in a mounted case (for ease of rotating each model) of lobster crustacyanin with and without astaxanthin

A laptop with molecular graphics examples

Molecular model of the DNA double helix set vertically on a plinth and in a plastic case but easily extractable for people to point with fingers at the base pairs

A poster of the mouse genome given as a pullout of an issue of Nature (on display on the back wall in Figure 1.1c) became a topical context for many people not least when the human genome was eventually first sequenced.

It is also important to look smart in my view. Also my tie would be carefully chosen to illustrate, for example, a regular pattern such as a piece of Escher art work or packing of spheres (the example in this photo of what I am wearing) or crystals.

Additional choices of exhibits would include other molecular models such as: a full molecular model of hen egg white lysozyme with bound hexasaccharide ('Labquip' model type assembled by my student Gail Bradbrook), a Beevers molecular model of a haemoglobin tetramer and a concanavalin A tetramer.

(a)

(b)

FIGURE 1.1 The University of Manchester School of Chemistry Open Days included a crystallography display. (a–d) show different views of such an exhibit; for details and names of staff involved see text.

(Continued)

(c)

(d)

FIGURE 1.1 (*Continued*)   (**See colour insert.**) The University of Manchester School of Chemistry Open Days included a crystallography display. (a–d) show different views of such an exhibit; for details and names of staff involved see text.

I would always start a discussion with a visitor to our display with either 'have you seen the double image seen through a calcite crystal?' or 'have you seen the DNA double helix?' These would be unfailing in offering a keen and friendly opener for people who may be quite shy.

1.2   UNIVERSITY OF MANCHESTER LECTURES TO SCHOOLS

There is a considerable outreach effort to bring the best science to schools and of course highlight the quality of the University of Manchester's science in particular. There are plenty of chances on offer to assist with these. As an example, I reproduce below my abstract for Fleetwood Grammar School near Blackpool (a private school, known in the United Kingdom for historical reasons as a public school).

Abstract: ***The Fascination of Crystals***
Prof John R Helliwell DSc, School of Chemistry, University of Manchester.

Summary
This talk will consider the nature and importance of crystals. Crystals allow the determination of precise molecular structures using X-rays and neutrons, which is vital for molecular structure and function studies. Thus new pharmaceuticals and functional materials can be designed. Crystals can also have specific properties, such as optical, magnetic and electric, which lead to interesting curiosities like the double refraction optical effect in crystals of calcite, based on the structural chemistry of the planar carbonate chemical group, as well as the exceedingly useful doped silicon crystal semiconductors that form the basis of modern computers. Closely allied to the study of crystals in molecular 3D structure determination are the solid state fibres and which for example yielded the DNA double helix 3D structure, the basis of genetic information storage, determined by X-ray fibre diffraction. This needed a clear mathematical theory for X-ray diffraction from a helix and led to the famous moment when Jim Watson recognised this effect in Rosalind Franklin's DNA fibre diffraction patterns. These examples will be illustrated with reference to a wide range of molecular models and crystals. The talk will conclude with an overview of the sophisticated technology of modern day synchrotron and neutron beams used these days to probe these states of matter.

It is worth recalling that in the first module of A Level Chemistry, the syllabus does cover states of matter and students are for example expected to:

* be able to explain the energy changes associated with changes of state;

* recognise the four types of crystal: ionic, metallic, giant covalent (macromolecular) and molecular;

* know the structures of the following crystals: sodium chloride, magnesium, diamond, graphite, iodine and ice;

* be able to relate the physical properties of materials to the type of structure and bonding present.

Overall then this talk will enrich, and move beyond, the Chemistry A Level curriculum and include undergraduate-level Chemistry concepts and theories through to modern day research and use of crystals. This final overview and summary of the talk seeks then to connect with expectations among the teachers and pupils/students regarding crystallography. In addition the talk will bring out the obvious inter-disciplinarity of crystallography reaching across physics, mathematics, chemistry and biology.

1.3   A MIXED AUDIENCE OF EXPERIENCED SCIENTISTS AS WELL AS ARTS AND HUMANITIES, WELL-EDUCATED PEOPLE

My University of Manchester 150th Anniversary Lecture, advertised as the W L Bragg Lecture, that I presented in 2001 in the University's Schuster (Physics) Department W L Bragg Lecture theatre was to an audience of experienced scientists in the University and, more widely, to members of the Manchester Literary and Philosophical Society. This can be viewed in full or just the lecture demonstration portions on their own at http://www.iucr.org/education/teaching-resources/bragg-lecture-2001.

I also explained how I prepared my lecture in my article:

Helliwell, J. R. (2009). *J. Appl. Cryst.* **42**, 365, doi: 10.1107/S0021889809002775.

This event included a sherry reception beforehand. I prepared a suite of well-labelled exhibits (including those listed in Section 1.1). The captions are reproduced in Appendix 1.A of this chapter.

A similar composition audience of science, arts and humanities as well as business people was present when I delivered a Friday Evening Discourse at The Royal Institution ('the RI') on 'Why do lobsters change colour on cooking?'.

The invitation to me from the Director of the RI, Baroness Susan Greenfield, is quoted in the following (Figure 1.2).

The Royal Institution
of Great Britain

Professor John Helliwell
Department of Chemistry
The University of Manchester
Oxford Road
Manchester M13 9Pl

The Royal Institution of Great Britain
21 Albermarle Street
London W15 4Bs
Switchboard 020 7409 2992
Fax 020 7629 3569
Email ri@ri.ac.uk
Web www.rigb.org
registered charity number 227938

Friday, 12 September 2003

Dear Professor Helliwell

Re: Friday Evening Discourse invitation

I am writing to invite you formally, on behalf of Baroness Susan Greenfield, to give a Friday Evening Discourse at the Royal Institution on 19 March 2004. If you are able and willing to give a Discourse on this date, we would be grateful if you could send us a brief biography of about 60 words, a synopsis of about 150 words and some relevant, colourful images. We will need this by mid-October 2003, and we will use it to advertise your Discourse on our website and in our Spring programme of Events.

As you may know, the Friday Evening Discourses go back to 1826, having been initiated by Michael Faraday, and many famous scientists and others have expounded their work subsequently. The audience is composed of Members and their friends, corporate (industrial) and school subscribers and invited guests —amounting to several hundred people who are interested in science but are, for the most part, not professional scientists. The tradition is that the Discourse should be informal; we like to hear

(a)

(b)

FIGURE 1.2 **(See colour insert.)** My Royal Institution ('The RI') Friday Evening Discourse April 2004. 'Why does a lobster change colour on cooking?' (a) Here we are assembled in the RI Library (accompanying the drinks were various science exhibits on marine colouration and on crystal structure analysis). (b) With the RI Director Baroness Susan Greenfield and Dr Peter Zagalsky, great expert on marine colouration biochemistry. (Note the Director's carefully chosen colour of her evening dress!)

the lecturer talking about his/her work rather than reading a prepared address. It is also a tradition that the lecture should last for exactly one hour, and be lavishly illustrated, wherever possible, with demonstrations, experiments, films, slides etc. Our Theatre Manager, Mr Bipin Parmar (bipin@ri.ac.uk), and the lectures staff here would be pleased to give any help you may need in the preparation of lecture demonstrations. Do arrange to speak to Mr Parmar as you start to consider what material you might use.

There is an exhibition in the Library associated with the Discourse arranged by our Exhibitions Organiser, Mrs Irena McCabe (irenam@ri.ac.uk), that illustrates and expands some of the ideas presented in the lecture. The exhibition is open to Members before and after the lecturer, and provides an important adjunct to the evening. Mrs McCabe is happy to advise on the exhibition and, in turn, is helped greatly by suggestions and indeed, if possible, material from the lecturer, so I would encourage you to make contact with her at an early stage.

Friday Evening Discourses will be published on the Royal Institution's web site www.rigb.org. We would be grateful if you could send a copy of the Discourse text to the Events Co-ordinator as soon as possible after giving the Discourse (or beforehand, if this is more convenient). The Discourse will be audio taped, and you are welcome to a copy of this tape if you think it will help you with the preparation of your text. The text may be anything from the summary of say 3000 words to an almost verbatim account, although something between 8000 and 12 000 is preferable.*

Baroness Susan Greenfield will host a small dinner party for the lecturer and guest and we hope that you could come to this. If you have any special dietary requirements, we would be grateful if you could let us know in good time. We would also be happy to provide you with accommodation should you like to stay overnight.

Please don't hesitate to contact me if you need any further information.

I look forward to hearing from you.

Yours sincerely
Dr Gail Cardew

Head of Programmes
Fax: 020 7670 2920

E-mail: gail@ri.ac.uk

1.4 COMMUNITY CENTRE LECTURES: E.G. THE WILMSLOW GUILD

The Wilmslow Guild was founded in 1926 and it continues to fulfil its original aims which are: 'To provide a centre in which men and women may find opportunities for the enrichment of life through education, fellowship and co-operative effect for the welfare of the community'.

https://www.wilmslowguild.org/

---

* I did write this up, but several years later following a talk I again gave on the topic at the Manchester Literary and Philosophical Society, published in their Memoirs series, and reproduced and somewhat extended in my *Crystallography Reviews* article reproduced in this book (Chapter 6).

SCIENCE MATTERS: A NEW SERIES OF SCIENCE LECTURES

Various Lecturers   *Class size 55*

The fourth series of science and technology lectures with brand new lectures which are thought provoking, exciting and entertaining, covering leading edge science and technology. Presented by experienced and enthusiastic lecturers who are mostly new to the Guild, the lectures will be of interest to both scientists and non-scientists alike.

21st September	Physics & Chemistry-more magic than Harry Potter	Dr Andrew G. Thomas
28th September	Making Sense of the Brain	Dr Rochelle Ackerley
5th October	The Gamblers Tale-randomness, chaos and order	Professor David Brooomhead
12th October	The Science of Climate Change: Current State of Knowledge and Challenges for the Future	Professor Hugh Coe
19th October	Wave Energy & the Manchester Botter	Professor Peter K Stansby
2nd November	Fysics of Frisbees 'n Further Flying Fings	Professor David Abrahams
9th November	The Molecules of Life & the Fascination of Crystals	Professor John Helliwell
16th November	Developments in Computer Science	Dr Adrian Jackson
23rd November	On the Origin of Species by Natural Selection	Dr Robert Callow
30th November	Einstein's Theory of Relativity	Professor Jeff Forshaw
Mondays 7:30–9:30 p.m.		Starting 21st September
Ten meetings		

Quoting Keith Wright <keith123wright@tiscali.co.uk>:

Dear John,

Just to say thanks once again for your excellent lecture last night. I was down at The Guild earlier today and many people told me how much they enjoyed it. Your efforts are much appreciated.

John Spawton, the Principal, will also write to you with his thanks and send a cheque covering your fee.

Thanks once again and I hope we can look forward to more presentations of yours at The Guild in the future.

Best Regards

Keith

1.5 INVITATION TO PRISONS VIA THE PRISONERS' EDUCATIONAL TRUST NEWSLETTER (WITHIN THE IYCr)

Trying to be imaginative, an as yet unfulfilled idea I have had was (is) to take crystallography into prisons as part of my contributions to the IYCr. My idea was that, whilst of course those in prison are there to pay their debt to society and their victims, another aspect has to be reform and self-improvement to turn prisoners of today into the good citizens of tomorrow. My idea to try and do this was prompted when the BBC Radio 4 Today programme had a piece about mentoring of prisoners. I had undertaken a variety of mentoring in the University (Manchester Gold scheme) and within the School of Chemistry as Senior Mentor for New Academics. This was of course no qualification for mentoring of prisoners! But, having undertaken numerous public understanding of science, engineering and technology lectures surely I could achieve some good here. Via the web, I found the Prisoners' Educational Trust who were helpful to carry my offer via their Newsletter (see the following figure). I have to admit that in my conversations with them, they were sceptical as they focussed on skills training such as prisoners learning car mechanics or being a chef.

See http://www.prisonerseducation.org.uk/newsletter (accessed 2 August 2015).

Did you know that 2014 is the UN- and UNESCO-endorsed International Year of Crystallography?

See http://iycr2014.org/

What is a crystal? How does crystallography figure in modern science? Why has the UN and UNESCO made 2014 the year of crystallography? Do you need to know crystallography if you wish to become a chef, a lab technician, a teacher, an MP, etc.? Modern crystal structure research gives a clear atomic level insight into genetics, that is from knowing the 3D structure of the DNA double helix, arguably the most important scientific advance of the twentieth century, or how anti-cancer agents work, or how computers depend on crystals, or how the future promise of Nanomaterials needs crystallography insights, etc. We have even explained 'Why do lobsters change colour on cooking!' Also our research tools include X-ray beams but how do we make and use X-rays for crystal structure analysis today? If you would like to learn more, during the IYCr 2014 and beyond, contact myself, Professor John R Helliwell DSc, john.helliwell@manchester.ac.uk.

For an example of one of my own outreach lectures, see:

http://www.iucr.org/education/teaching-resources/bragg-lecture-2001

Note: This was a special occasion, the 150th Anniversary of the Victoria University of Manchester, but it will give Prisons a feel for the discipline of crystallography.

A professional audience lecture that I gave has the slides available at:

http://www.iucr.org/education/teaching-resources/lonsdale-lecture-2011

These weblinks aim to give the reader further understanding for the subject matter and which covers physics, biology and chemistry.

1.6 NEWSLETTERS AND SYNCHROTRON FACILITY REPORTS

Example: The ALBA Spanish synchrotron radiation source February 2014 Newsletter, pp. 11–12. See http://issuu.com/albasynchrotron/docs/alba_february_def (accessed 2 August 2015).

Photo taken by John R Helliwell in his garden.

The inner life of plants and their responses

Have you ever wondered why plants in an identical pot and under obviously identical weather conditions respond differently? This photo (above) from my own garden in mid-February shows one part of a plant pot that has given up and one part that is still green and obviously alive. Have you ever read the Day of the Triffids by John Wyndham with the giant plants that take over the world? Or have you considered these ideas about life on Mars and beyond in potentially different atmospheres and soils than our own Earth?

● Plant response factors are the biochemical details to address such questions. Recent research published in Cell involves the auxin response. Auxin was originally discovered in research started by Charles Darwin and his son Francis looking at how plant growth responds to the direction of the light illumination (http://en.wikipedia.org/wiki/Auxin#Discovery_of_auxin).

Indole-3-acetic acid is the most abundant and basic auxin in plants. Auxin acts via regulation of genes in the plant. In the study by Boer et al published in Cell [1] they made a truly comprehensive study involving biophysical and biochemical characterisation techniques, including X-ray crystallography data quite recently measured at ALBA on the Xaloc beamline, as well as earlier data measured at ESRF in Grenoble before ALBA came online. From the 3D protein and nucleic acid structures the critical amino acids involved in the gene regulation could be identified. Modern genetics allows specific amino acids to be changed. The modified protein can be isolated and crystallised for further X-ray crystal structure analysis. The mutant G245A (glycine changed to alanine at position 245 in the particular protein polypeptide chain studied) was the one studied at ALBA. Earlier studies involved mutations at several other key places of the protein. These changes were deliberately introduced to disrupt or distort the protein nucleic acid interaction.

Prof John R Helliwell DSc
School of Chemistry, University of Manchester, UK
john.helliwell@manchester.ac.uk

With these site specific molecular changes in the genes the 'genetically modified plant' of the Arabidopsis (a small flowering plant related to cabbage and mustard) could be grown. These showed a variety of poor or compact (ie bushy) growing features.

How might this fundamental science make its way to impacting on society at large? Of course, as it did with this writer, it fuelled the imagination from arousing my renewed curiosity of what is happening in my own garden and on to my wilder imagination as to its implications in 'astrobiology'. Genetically modified (GM) crops are the more 'bread and butter' aspects. These are welcomed in some countries e.g. where hunger and famine are common place and controversial in others, notably where the citizens are well fed. How might this research help both community factions? For the hungry in the world such fundamental research will surely assist a more penetrating set of ideas and discoveries as to how to work with plant growers and agriculturalists as well as ultimately farmers. For the GM sceptics such research shows explicitly and clearly how plants respond to their environment via their genes and their biochemical response molecules, in this case auxin. Thus the wish by the sceptics of GM for a greater clarity on the effect of a given genetic mutation within the plant is achieved.

For myself I think the impact of this work is several fold. The team involved in the publication is clearly broad and each team is at the forefront of such modern day research; I am full of admiration. Secondly, as Chairman of the ALBA Beamtime Panel and of the Science Advisory Committee, for me to be able to see over a 4 years period of time the ALBA facility in general and Xaloc in particular move from build to commissioning to regular use is a marvellous thing. The Spanish community in all its range of science and technology skills can be rightly proud of this achievement with ALBA. Thirdly I have learnt a lot more than I did about plant molecular biology. Fourthly, maybe it is time for me to start that novel as a modern successor to the 'Day of the Triffids'!

[1] Boer et al Cell Volume 156, Issue 3, 577-589, 30 January 2014.

Science in detail

FEBRUARY 2014 - No. 37 | *ALBA news* | 11

If you want to know more about this experiment, go to page 15.

1.A APPENDIX

I prepared an Exhibition to accompany my University of Manchester W L Bragg Lecture of 2001; the captions for the exhibits are given below.

1.A.1 THE DIFFERENCE BETWEEN DIFFRACTION AND MICROSCOPY EXPLAINED

In a visible light microscope, the sample is illuminated and a glass lens combines the transmitted light to produce a magnified image. The microscope resolves detail at the level of cells and larger-scale objects. The resolvable detail cannot be finer than the wavelength/2 of the illuminating light. Visible light has a wavelength of around 500 nm.

Can we have an X-ray microscope to directly resolve details at the atomic level (spacings of 0.1 nm or so)? This would need X-ray lenses equivalent to the lenses in a visible light microscope. These do not exist. So, we must perform the lens function mathematically, that is via the computer. The analysis is known as Fourier series, named after the French mathematician Jean Baptiste Joseph Fourier (1768–1830) who first introduced this mathematical method.

1.A.2   Lysozyme Enzyme X-Ray Crystal Structure

'All atom' model, excluding hydrogens, of this globular protein was determined with X-ray crystal-lography by D C Phillips and co-workers in the early 1960s at The Royal Institution in London, and where Sir W L Bragg was Director (1953–1966). Lysozyme itself was discovered by Sir Alexander Fleming in 1922. In green is the hexasaccharide substrate, which is a model for the cell wall poly-saccharide of a bacterium. This enzyme is found in hen egg white, for example, and serves as an anti-bacterial molecular defence. The aspartic acid residue 52 and glutamic acid residue 35 sit on either side of the bond linking sugar sites 'D' and 'E' (see cotton threads linked to the key bond). The bond is strained and then broken by these enzyme active site residues. Thus, the bacterial cell wall is punctured and bursts. The bacterium and the protein then disassociate and the enzyme is 'reprimed' by the addition of a water derived proton at the glu 35 carboxyl group, ready for the next bacterium. Lysozyme is one of Nature's catalysts. This model was built by Gail Bradbrook.

See D C Phillips (1966) *Scientific American* 215(5): 78–90.

1.A.3   Concanavalin A Protein Tetramer X-Ray Crystal Structure

'Beevers' type protein model of concanavalin A involving one bead per amino acid residue. This pro-tein is an association of four monomers each of 237 amino acids. The protein binds sugar molecules but is not an enzyme. Instead, it serves as a scaffold, cross-linking protein to sugar to protein, etc., providing vegetable beans with an anti-fungus defence. It is a commonly occurring protein found both in large quantities in jack beans and in different types of beans. The family of proteins is known as the legume lectins. Concanavalin A is the most common member. It is also widely used as a model system for understanding the molecular biophysics and biophysical chemistry of protein ligand interactions and energetics. This protein was first crystallized in 1919 by the Nobel prize winner J B Sumner in Corenll, United States. The X-ray crystal structure was first determined by three groups (two from the United States and one from Israel in the 1970s). Since then, its structure has been extensively studied by the Helliwell lab via synchrotron radiation and neutron protein crystallography as well as by molec-ular dynamics (Gail Bradbrook), in both saccharide-free and glucose- or mannose-bound forms. Data collection was undertaken either at the Daresbury SRS or the Cornell 'CHESS' Synchrotron (United States), and the neutron facility at the Institut Laue Langevin, Grenoble respectively.

1.A.4   Hydroxymethylbilane Synthase Enzyme X-Ray Crystal Structure in its Active Form

This enzyme is responsible for catalyzing the polymerisation of pyrrole units to form tetrapyrrole, which after release from the enzyme, can cyclize and this serves as the precursor of haem, vitamin B12 or chlorophyll via incorporation of an atom of iron, cobalt or magnesium, respectively. Absence of this enzyme at the genetic level causes the malady known as porphyria (the madness of King George). The enzyme crystal structure in its oxidized form was determined by the laboratory of Sir T L Blundell in Birkbeck College, University of London. This 'Beevers' model (one bead = one amino acid) shows the enzyme in its reduced, that is active form. (The cofactor being the active agent, shown in orange, atom by atom.) This was determined by the Helliwell lab, in collaboration with Dr Alfonse Haedener of the University of Basle, Switzerland, using the Daresbury Synchrotron Radiation Source and the Multi-wavelength Anomalous Dispersion (MAD) technique. In addition, the white extended piece embedded near the cofactor represents the growth of the electron density when an enzyme crystal has been fed substrate and the time-resolved crystal structures have been determined using Laue diffraction (shown here is the '2 hour' electron density); this data collection was done at the European Synchrotron Radiation Facility, Grenoble, France.

1.A.5 'Beevers Lipson Strips'

Before the advent of computers, the calculation of the Fourier Series transforms of the diffraction data to obtain electron density contour maps was undertaken with the help of these boxes of cosine and sine compilations. A calculation of one projection image alone could take many months. Professor Henry Lipson FRS (1910–1991), who had been a co-worker of Sir W L Bragg, was a Head of the UMIST Physics Department, a past Chairman of the Manchester Branch of the Institute of Physics and a past President of the Manchester Literary and Philosophical Society. Dr Arnold Beevers (1908–2001), who was based in Liverpool, joined with Lipson at Manchester in this research and development.

1.A.6 The DNA Double Helix Structure

Through a combination of model building, DNA helical X-ray diffraction patterns and previously available chemical composition data, the structure of the DNA double helix was deduced by James D Watson and Francis Crick at the Cavendish Laboratory, Cambridge. The work was published in *Nature* in 1953. The Director of the Cavendish Laboratory during the period of the work was Sir W L Bragg. The fibre diffraction experimental data were recorded by Rosalind Franklin, in conjunction with Maurice Wilkins at Kings College, University of London.

With the hydrogen bonding complementarity between the nucleotide bases on the inside of each strand, the structure immediately suggested a way that genetic inheritance could be passed from the parents to a child. This scientific result has been described as the greatest single scientific achievement of the twentieth century.

Section II

Celebrating the centennial of
X-ray crystal structure analysis

2

The centennial of the first X-ray crystal structures*

John R. Helliwell[†]

School of Chemistry, University of Manchester, Manchester M13 9PL, UK

(Received 30 August 2012; final version received 12 September 2012)

A short account is given of the discoveries of the first crystal structures using X-ray diffraction by William Henry Bragg and William Lawrence Bragg, a father and son team working at Leeds University and Cambridge University. Their first publications, separately and together, are highlighted. This is a contribution to the centennial celebrations of these discoveries and looking towards the International Year of Crystallography to be celebrated in 2014 led by the International Union of Crystallography.

Keywords: William Henry Bragg; William Lawrence Bragg; first X-ray crystal structures of W.L. Bragg; X-ray spectrometer of W.H. Bragg

Contents

* From *Crystallography Reviews,* Vol. 18, No. 4, October 2012, 280–297.
† Email: john.helliwell@manchester.ac.uk

2.1 Introduction

X-ray crystal structure analysis, and its development, was instigated 100 years ago. The nature of X-rays as waves or corpuscles was a controversy and the thinking on the nature of the electron distributions in an atom was before quantum mechanics. The structures of molecules were undetermined. William Henry Bragg (WHB; 1862–1942) and William Lawrence Bragg (WLB; 1890–1971), father and son, played the pivotal roles at Leeds University and Cambridge University, in pioneering X-ray crystal structure analysis through 1912–1914, interrupted by the eruption of World War I (WWI) in Europe in August 1914. The Braggs had promptly built upon the first 'X-ray diffraction from a crystal' set of experiments undertaken in Munich by Max von Laue, Walter Friedrich and Paul Knipping, with key roles by Paul Ewald and Arnold Sommerfeld in early 1912. This proved conclusively that X-rays as waves were diffracted by the crystal as a 3-D diffraction grating. Today, we refer to the number of X-ray photons per second incident onto our crystal and our crystal diffracts the X-rays, now as waves. We live with 'wave-particle duality' as a common place. Strong evidence for X-rays as waves came from Charles Glover Barkla's research at Liverpool University through the 1900s on the polarization of X-rays. There were direct clashes in the science literature of the time involving Barkla at Liverpool and WHB, then at Adelaide University, Australia, as he, unlike Barkla, was convinced that X-rays were corpuscular by nature. Nobel Prizes in Physics were awarded to Laue (1914), both the Braggs (1915) and Barkla (1917). The X-ray spectroscopy of the elements by Henry ('Harry') Moseley in Manchester was tragically ended when Moseley, by then at the war front in Gallipoli, was killed in 1915. WHB and Earnest[1] Rutherford were regularly in touch by correspondence. This commenced when WHB was in Adelaide and Rutherford was in Canada. Their friendship continued when they were both in the North of England, WHB in Leeds and Rutherford in Manchester. WHB also served as External Examiner for Manchester University Physics at Rutherford's invitation. C.G. Darwin, also in Manchester Physics, grandson of Charles Darwin, derived a key equation for the diffraction of X-rays by a crystal (in 1914).

Various celebrations are being held to mark the importance of the first crystal structures. In late 2012, the Asian Crystallographic Association meeting was held in Adelaide, because of the WHB work there over a more than 20-year period, mainly physics teaching but building up research on radioactivity and the nature of X-rays. The research into the stopping distance of alpha particles in matter by WHB in Adelaide, and its medical potential, was arguably the most important. The European Crystallographic Meeting (ECM28) 2013 was hosted by the United Kingdom at Warwick University and had a special centennial celebration session. Recently in March 2012, there was a special celebration conference held in Munich by the German Crystallographic Society and a special lecture was presented by Prof. Dr Dieter Schwarzenbach at ECM27 held in August 2012 in Bergen.

2.2 WHB and WLB; the father and son team

WHB and WLB developed, over many decades, X-ray crystal structure analysis. WHB was based at University College London (UCL; from 1915) and then at the Royal Institution (RI; 1923–1935). WLB became based at Manchester University (1919–1937), then at the Cavendish Laboratory in Cambridge (1938–1953) and finally at the RI (up to 1965). Of those initial years 1912–1915, in the modern era, we refer back specifically to 'Laue diffraction', to Bragg's Law (WLB 1912 in Cambridge), the first crystal structures, which were of the alkali halides (WLB 1913 in Cambridge using Laue diffraction photos, and also in Leeds using his father's diffractometer, see W.L. Bragg (*1*)), the first X-ray diffractometer (WHB (*2*)) and Fourier analysis (WHB at UCL in 1915).

The sequence of events above is described in various biographical works, including by Gwendolen Caroe on her father, WHB (*3–5*). There is an extensive collection of both the Braggs' archives held at the RI. These formed the touchstone for the biographical books. The nature of the father and son work and relationship is described particularly in the recent book by Jenkin (*5*) and disagrees with some interpretations in Hunter (*4*). That WHB spoke highly of his son, WLB, and that WLB admired his father, is abundantly clear. However, the question of the recognition of WLB's work independent of his father remained and is a matter of much written analysis. Recall that WLB was a Cambridge first year postgraduate, when the inspiration for his 'Bragg's Law' struck his mind whilst 'walking on the Cambridge backs'. WHB was an established Professor of Physics at Leeds University. It was WHB who attended the 1913 Solvay Conference on Physics, held in October 1913, without WLB, but WLB received a postcard signed by Sommerfeld, Marie Curie, Laue, Einstein, Lorentz, Rutherford and others congratulating him for 'advancing the course of natural science' (*5*). WHB and WLB are the only father and son joint recipients of the Nobel Prize for Physics and WLB is still the youngest recipient.

WHB was born in Cumbria in 1862, and when his mother died when he was only 7 years old he was moved to Market Harborough to live with the family of his uncle, also called William. WHB was educated as a boarder at the King William's College in the Isle of Man: (a fact commemorated on an Isle of Man stamp). He went from there to Trinity College, Cambridge, to study mathematics, graduating as a 'wrangler' ahead of the mathematician Whitehead (*5*). J.J. Thompson, Head of the Cavendish Laboratory, proved influential in WHB's appointment as a very young Professor of Maths and Physics at Adelaide University, 'learning physics on the boat on the way out to Australia'. WHB was 23 years there, married and had a family of two sons and a daughter. When WLB, the eldest son, was nearly 16 years old, he entered Adelaide University to study, mainly, maths and physics, but also chemistry, and was 18 years old when WHB accepted the appointment as Chair of Physics at Leeds University in 1908, 'in order to be more at the centre of things' (*5*). This included WHB being nearer to his scientific friend and correspondent, Rutherford, who was by then based in Manchester. As for WLB, he, like his father had, entered Trinity College Cambridge in 1909 and graduated with first class honours in physics in 1911 (*5*).

2.3 How did WHB get to know of the Munich X-ray diffraction experiment?

Jenkin (*5*) states that 'not highlighted before, there was a letter sent by a Norwegian, Lars Vegard to WHB who knew of WHB's strong interest in the nature of X-rays, and who included a copy, with Laue's blessing, of an X-ray diffraction photograph. Vegard also explained various details of the work'. Father and son discussed the work in detail, including during their summer holiday in 1912, which was at Cloughton, near Scarborough, UK (*5*).

2.4 Their first experiments

WLB himself makes clear the sequence of events in his own lifetime look-back book *The Development of X-ray Analysis* (*6*) involving repeating the 'Laue diffraction experiment' but on crystals of NaCl, KCl, KBr and KI in the Cavendish Laboratory and repeating the work on his father's X-ray spectrometer, which was 'more powerful'. WLB's paper in 1913 (*1*) is under his own name, WLB, (i.e. without a doctorate at that point), includes the raw data comprising the 'Laue diffraction' photos (modern terminology) and also some of the X-ray spectrometer scans. The paper (*1*) is written without an address, is communicated by his father and is received at The Royal Society on 21 June 1913 and read on 27 June 1913. WHB's 'X-ray spectrometer' is under his name only, i.e. not involving WLB, and is described in *Nature* (*2*). There are various joint papers in this period; I will quote from one in particular (*7*). This was 'received on April 7, 1913 and read on April 17, 1913'. WHB's address is the Department of Physics, University of Leeds and WLB's was Trinity College, Cambridge University. The article's first reference is to WLB presenting the interpretation of Laue diffraction photographs by means of 'reflection of X-rays in such planes within the crystal as are rich in atoms' (at the Cambridge Philosophical Society, 11 November 1912). Their figure (Figure 2.1 here) is reproduced with permission of the Royal Society (but who point out that articles of that age are by now out of copyright). On page 436 is the footnote that 'We learn that Messrs.

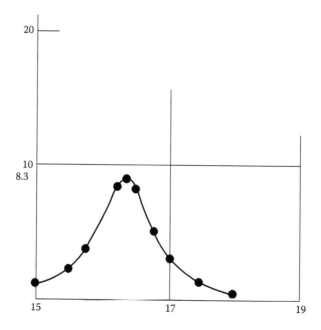

Figure 2.1 Regular reflection from cleavage face of rock salt, glancing angle 8.3°. [This is the original article's figure caption (see ref 7).]

Moseley and Darwin have lately been making similar experiments to some of those recorded here. Their results, which agree with ours, have not been published'. The end of the article in its concluding paragraph includes the statement: 'The effect which we have been describing is clearly identical in part with that which Prof. Barkla (using an X-ray sensitive photographic plate) has described ... in an abstract. But it seems probable that the ionization method can follow the details of the effect more closely than the photographic method has so far been able to do'. Evidently there was a close running competition between Leeds, Manchester, and Liverpool Universities.

Figure 2.2 shows the layout of NaCl derived by WLB (*1*). Their sequence of publications is illustrated in Figure 2.3 as simple snapshots of the title and opening words along with one page of the experimental notebook of WHB from July 1913, featuring the raw data measurements on diamond X-ray diffraction. These articles tell a fascinating and unfolding scientific story.

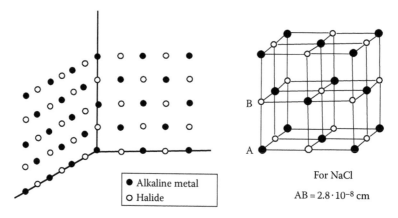

Figure 2.2 See W.L. Bragg (*1*). [This is the original article's figure caption (see ref 1).]

2.5 After the first crystal structures

As the anxieties about a war with Germany grew, WHB and WLB remained focussed on their discoveries until WLB entered the British Army after the breakout of the war in August 1914. Thus, for example, WHB published X-ray diffraction data on crystalline sulphur and quartz (submitted in December 1913 and published in March 1914; Figure 2.4) and for whose crystal structures a solution could not be immediately obtained. The title refers to the X-ray spectrometer, WHB's preferred name for his, what we would call today, X-ray diffractometer; the point being that WHB also measured X-ray spectra with it. The alpha quartz crystal structure was later established as being in space group $P3_121$ and the molecular structure having linked SiO_4 units. Quartz also has other polymorphs.

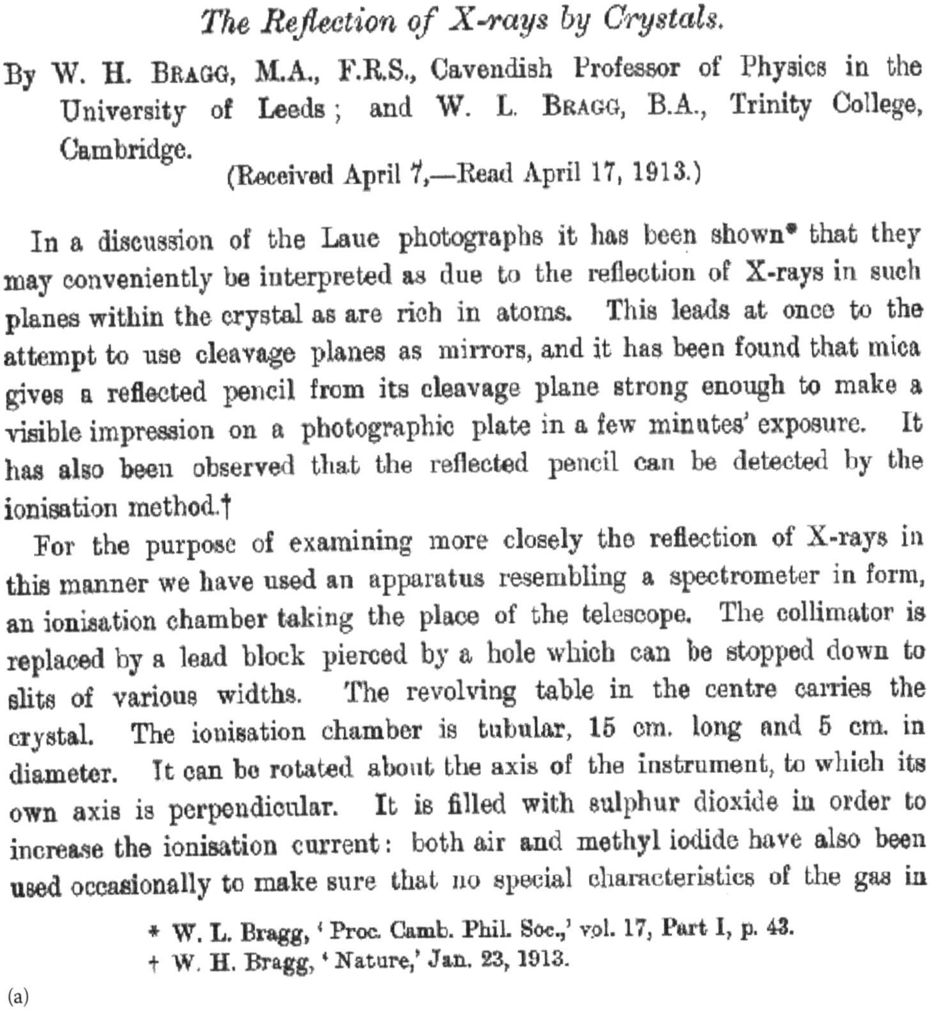

The Reflection of X-rays by Crystals.

By W. H. BRAGG, M.A., F.R.S., Cavendish Professor of Physics in the University of Leeds; and W. L. BRAGG, B.A., Trinity College, Cambridge.

(Received April 7,—Read April 17, 1913.)

In a discussion of the Laue photographs it has been shown[*] that they may conveniently be interpreted as due to the reflection of X-rays in such planes within the crystal as are rich in atoms. This leads at once to the attempt to use cleavage planes as mirrors, and it has been found that mica gives a reflected pencil from its cleavage plane strong enough to make a visible impression on a photographic plate in a few minutes' exposure. It has also been observed that the reflected pencil can be detected by the ionisation method.[†]

For the purpose of examining more closely the reflection of X-rays in this manner we have used an apparatus resembling a spectrometer in form, an ionisation chamber taking the place of the telescope. The collimator is replaced by a lead block pierced by a hole which can be stopped down to slits of various widths. The revolving table in the centre carries the crystal. The ionisation chamber is tubular, 15 cm. long and 5 cm. in diameter. It can be rotated about the axis of the instrument, to which its own axis is perpendicular. It is filled with sulphur dioxide in order to increase the ionisation current: both air and methyl iodide have also been used occasionally to make sure that no special characteristics of the gas in

[*] W. L. Bragg, 'Proc. Camb. Phil. Soc.,' vol. 17, Part I, p. 43.
[†] W. H. Bragg, 'Nature,' Jan. 23, 1913.

(a)

Figure 2.3 A collection of snapshots of the sequence of publications, together and separately, by WHB and WLB. From: (a) W.H. Bragg and W.L. Bragg (7). *(Continued)*

The Reflection of X-rays by Crystals. (II.)

By W. H. BRAGG, M.A., F.R.S., Cavendish Professor of Physics in the University of Leeds.

(Received June 21,—Read June 26, 1913.

This note is a supplement to a paper on the reflection of X-rays by crystals which has been recently communicated to the Royal Society.* It is there shown that the wave-length of a homogeneous beam of X-rays can be found accurately in terms of the spacing of the elements of a crystal. There has been some doubt as to the actual arrangement of the atoms in the crystal and in consequence it was not possible in the paper quoted to draw any final conclusions as to wave-length values. From the work now described by W. L. Bragg it appears that the reflection phenomena lead to a more definite knowledge of crystal structure, and we may now complete various quantitative determinations.

The elementary volume in rock-salt is a cube with 1 atom of sodium at each of four corners and 1 atom of chlorine at each of the other four. In other words the number of elementary volumes in any space of measurable dimensions is equal to the number of atoms in that space.

The number of molecules in 1 c.c. of NaCl is

$$2 \cdot 15 / 58 \cdot 5 \times 1 \cdot 64 \times 10^{-24} = 2 \cdot 24 \times 10^{22}$$

(The weight of the H atom is taken to be $1 \cdot 64 \times 10^{-24}$.)

The number of atoms is twice as great and the elementary cube volume is therefore $1 / 4 \cdot 48 \times 10^{22} = 2 \cdot 23 \times 10^{-23}$. The edge of the cube is $2 \cdot 81 \times 10^{-8}$; this is the distance between consecutive reflecting planes parallel to (100).

The principal bundle of homogeneous X-rays from a platinum anticathode is stated in the paper quoted to be reflected at the (100) face of rock-salt at a glancing angle of $11 \cdot 55°$. Recent observations with better apparatus show that this bundle is really double, consisting of two separate sets whose wave-lengths differ from each other by a little less than 2 per cent. of either; they also show that the first estimate was a little too high. For the purpose of the present argument it is sufficiently accurate to ignore the division and assume the angle to be $11 \cdot 3°$. This gives a wave-length

$$(2 \, d \sin \theta) = 2 \times 2 \cdot 81 \times 10^{-8} \times 0 \cdot 196 = 1 \cdot 10 \times 10^{-8}$$

The wave-lengths of other homogeneous rays can then be found easily as soon as their angles of reflection are known.

* W. H. Bragg and W. L. Bragg, these 'Proceedings,' A, vol. 88, p. 428.

(b)

Figure 2.3 (*Continued*) From: (b) W.H. Bragg (8).   (*Continued*)

The Structure of Some Crystals as Indicated by their Diffraction of X-rays.

By W. L. BRAGG, B.A.

(Communicated by Prof. W. H. Bragg, F.R.S.   Received June 21,—Read June 26, 1913.)

[PLATE 10.]

A new method of investigating the structure of a crystal has been afforded by the work of Laue[*] and his collaborators on the diffraction of X-rays by crystals. The phenomena which they were the first to investigate, and which have since been observed by many others, lend themselves readily to the explanation proposed by Laue, who supposed that electromagnetic waves of very short wave-lengths were diffracted by a set of small obstacles arranged on a regular point system in space. In analysing the interference pattern obtained with a zincblende crystal, Laue, in his original memoir, came to the conclusion that the primary radiation possessed a spectrum consisting of narrow bands, in fact, that it was composed of a series of six or seven approximately homogeneous wave trains.

In a recent paper[†] I tried to show that the need for assuming this complexity was avoided by the adoption of a point system for the cubic crystal of zincblende which differed from the system considered by Laue. I supposed the diffracting centres to be arranged in a simple cubic space lattice, the element of the pattern being a cube with a point at each corner, and one at the centre of each cube face. A simpler conception of the radiation then became possible. It might be looked on as continuous over a wide range of wave-lengths, or as a series of independent pulses, and there was no longer any need to assume the existence of lines or narrow bands in its spectrum.

[*] W. Friedrich, P. Knipping, and M. Laue, 'Münch. Ber.,' June, 1912.
[†] 'Camb. Phil. Soc. Proc.,' November, 1912.

(c)

Figure 2.3 (*Continued*) From: (c) W.L. Bragg (*1*).   (*Continued*)

The Structure of the Diamond.

By W. H. Bragg, M.A., F.R.S., Cavendish Professor of Physics in the University of Leeds, and W. L. Bragg, B.A., Trinity College, Cambridge.

(Received July 30, 1913.)

There are two distinct methods by which the X-rays may be made to help to a determination of crystal structure. The first is based on the Laue photograph and implies the reference of each spot on the photograph to its proper reflecting plane within the crystal. It then yields information as to the positions of these planes and the relative numbers of atoms which they contain. The X-rays used are the heterogeneous rays which issue from certain bulbs, for example, from the commonly used bulb which contains a platinum anticathode.

The second method is based on the fact that homogeneous X-rays of wave-length λ are reflected from a set of parallel and similar crystal planes at an angle θ (and no other angle) when the relation $n\lambda = 2d\sin\theta$ is fulfilled. Here d is the distance between the successive planes, θ is the glancing angle which the incident and reflected rays make with the planes, and n is a whole number which in practice so far ranges from one to five. In this method the X-rays used are those homogeneous beams which issue in considerable intensity from some X-ray bulbs, and are characteristic radiations of the metal of the anticathode. Platinum, for example, emits several such beams in addition to the heterogeneous radiation already mentioned. A bulb having a rhodium anticathode, which was constructed in order to obtain a radiation having about half the wave-length of the platinum characteristic

---

278   Prof. W. H. Bragg and Mr. W. L. Bragg.

rays, has been found to give a very strong homogeneous radiation consisting of one main beam of wave-length 0.607×10^{-8} cm.*, and a much less intense beam of wave-length 0.533×10^{-8} cm. It gives relatively little heterogeneous radiation. Its spectrum, as given by the (100) planes of rock-salt, is shown in fig. 1. It is very convenient for the application of the second method. Bulbs having nickel, tungsten, or iridium anticathodes have not so far been found convenient; the former two because their homogeneous radiations are relatively weak, the last because it is of much the same

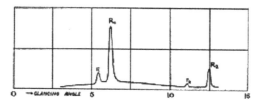

Fig. 1.—Spectra of rhodium rays : 100 planes of rock-salt.

I
(d)

Figure 2.3 (*Continued*) From: (d) W.H. Bragg and W.L. Bragg (9).

(*Continued*)

wave-length as the heterogeneous rays which the bulb emits, while it is well to have the two sets of rays quite distinct. The platinum homogeneous rays are of lengths somewhat greater than the average wave-length of the general heterogeneous radiation; the series of homogeneous iridium rays are very like the series of platinum rays raised one octave higher. For convenience, the two methods may be called the method of the Laue photograph, or, briefly, the photographic method, and the reflection method. The former requires heterogeneous rays, the latter homogeneous. The two methods throw light upon the subject from very different points and are mutually helpful.

The present paper is confined almost entirely to an account of the application of the two methods to an analysis of the structure of the diamond.

The diamond is a crystal which attracts investigation by the two new methods, because in the first place it contains only one kind of atom, and in the second its crystallographic properties indicate a fairly simple structure. We will consider, in the first place, the evidence given by the reflection method.

The diagram of fig. 2 shows the spectrum of the rhodium rays thrown by the (111) face, the natural cleavage face of the diamond. The method of obtaining such diagrams, and their interpretation, are given in a preceding

* This value is deduced from the positions of the spectra of the rhodium rays in the (100) planes of rock-salt on the assumption that the structure of rock-salt is as recently described (see preceding paper).

II

(d)

(e)

Figure 2.3 (*Continued*) (e) This publication's raw diffraction data can be found in the original experimental notebook held at Leeds University, United Kingdom (available on the web at URL: http://www.leeds.ac.uk/library/spcoll/bragg-notebook/pdf.htm). (*Continued*)

The Analysis of Crystals by the X-ray Spectrometer.

By W. Lawrence Bragg, B.A.

(Communicated by Prof. W. H. Bragg, F.R.S. Received November 13,—Read
November 27, 1913.)

In a former communication to the Royal Society,* an attempt was made
to determine for certain crystals the exact nature of the diffracting system
which produces the Laue X-ray diffraction photographs. The crystals chosen
for particular investigation were the isomorphous alkaline halides NaCl, KCl,
KBr, and KI. As in the original experiments of Laue and his collaborators,
a thin section of crystal was placed in the path of a narrow beam of X-rays,
and the radiation diffracted by the crystal made its impression on a
photographic plate. By noticing what differences were caused in the
photograph by the substitution of heavier for lighter atoms in the crystal, a
definite arrangement was decided on as that of the diffracting points of the
crystalline grating.

* W. L. Bragg, 'Roy. Soc. Proc.,' A, vol. 89, p. 249.

The Analysis of Crystals by the X-ray Spectrometer.   469

Though it was found possible in the case of these simple salts to determine
the position of the atoms of alkaline metal and halogen, which constitute the
elements of the dimensional diffraction grating, yet this method, which may
be called the photographic method, is very limited in its range of applications.
It was only the extremely simple nature of the NaCl structure which made
its analysis possible. On the other hand, the X-ray spectrometer, which has
been devised by W. H. Bragg for the purpose of studying the reflection of
X-rays by crystals,* affords a very much more powerful method of research
into the structure of the crystal.

The photographic method works by throwing on the crystal a beam of
"white" X-radiation, and comparing the strength of the beams reflected by
various types of planes (nets) of the point system on which the atoms are
arranged. The X-ray spectrometer employs a monochromatic radiation and
faces are examined in detail one by one. In the first place, the spectrometer
tells the distance in centimetres of plane from plane parallel to these faces.
Moreover, if the successive planes are of identical composition, the results of
the examination show this. If on the other hand the planes occur in groups,
each group containing several planes of different nature, it is hoped that the
results given below will show how the instrument can be made to give the
exact spacing and relative masses of the planes of these groups. This means
that we can obtain enough equations to solve the structure of any crystal,
however complicated, although the solution is not always easy to find. In
this paper I wish to indicate the solution for several types of crystals. For
many of the experimental results I am indebted to my father, the rest have
been obtained in Leeds with one of the spectrometers which he has
constructed.

Parallel to any one of its possible faces, a crystal may be regarded as being
built up of a series of planes. Each plane passes through the centres of one
or more sets of atoms identical in all respects. The successive planes,

(f)

Figure 2.3 (*Continued*) (f) W.L. Bragg (*10*).

****

X-RAY CRYSTALLOGRAPHY

The new knowledge of the atomic structure of matter uncovered
over the past half-century by the X-ray-diffraction technique
has led to a fundamental revision of ideas in many sciences

by Sir Lawrence Bragg

Fifty-six years ago a new branch of science was born with the discovery by Max von Laue of Germany that a beam of X rays could be diffracted, or scattered, in an orderly way by the orderly array of atoms in a crystal. At first the main interest in von Laue's dis-

summed up in a "model." I have often been asked: "Why are you always showing and talking about models? Other kinds of scientists do not do this." The answer is that what the investigator has been seeking all along is simply a structural plan, a map if you will, that shows

atoms in a crystal of sodium chloride (ordinary table salt) is 2.81 angstrom units (an angstrom is 10^{-10} meter), whereas the most commonly used wavelength in X-ray analysis is 1.54 angstroms.

Actually crystals came into the picture only because they are a convenient

(g)

Figure 2.3 (*Continued*) (g) W.L. Bragg (*11*).

The X-ray Spectra given by Crystals of Sulphur and Quartz.

By W. H. Bragg, M.A., F.R.S., Cavendish Professor of Physics in the
University of Leeds.

(Received December 1, 1913,—Read January 29, 1914.)

A number of results of examination of crystals by means of the X-ray spectrometer, which have been recently obtained in this laboratory, are discussed in a paper by W. L. Bragg.* The cases which he has considered are those in which the information has been sufficient for a complete solution of the crystal structure. This note deals with two cases, viz., sulphur and quartz, which have not been completely solved, but which have nevertheless given interesting results.

Figure 2.4 As war approached, WHB's attention turned towards those crystal structures that could not be immediately solved: e.g. the cases of crystalline sulphur and quartz. From W.H. Bragg (*12*).

2.6 After World War I (WWI)

2.6.1 *Science*

Their science work resumed. WHB at University College London concentrated on the X-ray crystal structures of organics, whilst WLB at Manchester University concentrated on the crystal structures of inorganics in general and silicates in particular. For WLB, this included the X-ray crystal structure of calcite and explaining the optical property of birefringence. This required the understanding of the absolute intensities based on X-ray atomic scattering factors, developed with Douglas Hartree, presumably both to the appreciation of his physics departmental colleagues. A developing feature of WLB's career was the suspicion from his physics department colleagues, first in Manchester and then at Cambridge, that X-ray crystallography was not 'proper physics'. (A referee for this article remarked 'It was interesting to learn that some physicists didn't regard all this 'crystal structure analysis stuff' as 'proper' physics. There's some chemists who don't regard crystallography as 'proper' chemistry either! Both of these attitudes smack of sour grapes, and crystallography is what it is: an absolute essential to a broad spectrum of modern science, and a discipline that has defined structural science in chemistry, biology, materials science, physics and elsewhere. People need to be reminded of its origins'.)

2.6.2 Administration of science

In this short article, many details, including very significant ones, are left out. The respective obituary notice and biographical memoirs of the Royal Society give separate comprehensive summaries of the lives and scientific outputs of WHB and WLB (13,14). Therein are descriptions of the details of the influential role of WHB in British science and society 'between the wars', as measured by his Order of Merit, one of only 24 persons at any one time selected by the ruling King or Queen. WHB also served as President of the Royal Society.

For WLB, there was his important role in WWI as a key 'science and technical person' from the British side, in association with the French, for developing sound ranging to pinpoint German gun emplacements. This led to his achieving the rank of Major and the award of the Military Cross (5). Jenkin (5) gives a précis of the written evidence which points to sound ranging being as important as the introduction of tanks by the allies in concluding the WWI. Alongside this was the family tragedy of the death of Bob Bragg, WLB's younger brother, basically in the same military operation as took the life of Harry Moseley. WHB was 52 when the death of Bob happened.

WLB went from Manchester, via a short time (a year) as Director of the National Physical Laboratory (all that administration was not for him) before succeeding Rutherford again, this time as Head of the Cavendish Laboratory in 1938. Under WLB's direction came the solution of the first protein crystal structures by Max Perutz and John Kendrew, as well as the double helix by Francis Crick and James Watson. This latter discovery also linked to the experimental DNA fibre diffraction work of Rosalind Franklin and Maurice Wilkins, who were based at King's College, London. During WLB's term as Director of the RI, David Phillips and colleagues solved the first enzyme crystal structure. I would observe that WLB had at the end of WWI been in charge of 40 sound ranging stations of 50 persons each. Thus, WLB's management and leadership, learnt in WWI, were also further developed and applied by him in these scientific roles as Director.

2.6.3 Taking science to the public and to school children

Both WHB and WLB were fine expositors of science to the public and to school children, especially through the Adelaide (WHB) and the RI periods (WHB and WLB), both giving sets of RI Christmas Lectures. They both evidently had a great way of explaining complex things simply and by analogy. An example for WHB is his book (15) The Universe of Light based on his RI Christmas Lectures of 1931, which is superbly illustrated. For WLB, see for example, his Scientific American article on X-ray crystallography (11). Joel Bernstein and I were teaching at a crystallography school in Como recently and he told me of hearing a lecture by WLB at Yale University in the 1960s; he vividly recalled that WLB described a crystal as a 'symphony of electrons', a beautiful thought.

2.7 Conclusions

Paraphrasing WLB (11), 'The new knowledge of the atomic structure of matter uncovered in the past century by the (X-ray) diffraction technique has led to a fundamental revision of ideas in many sciences'. To X-ray diffraction has been added electron and neutron diffraction. X-ray diffraction

Figure 2.5 WHB and WLB in 1915 (approximate estimate); from the Nobel Prize website.

is sensitive to the very finest details of electron density via the steady development and study of electron charge density distributions and spin and momentum densities in crystals are determined via neutron diffraction (*16*). The rigid limitation of X-ray diffraction to be 'the static technique' has been shown to be not always so across many time domains of measurement principally through harnessing synchrotron X-radiation (see e.g. *17*). Most recently, there is a major extension of such capabilities into the femtosecond time-resolution domain arrived at with the new X-ray lasers generation of sources. Being able 'to see atoms', as WLB vividly described X-ray crystal structure analysis, has found so many applications in different areas of science. In the spirit of this article being a historically oriented piece, it is worth mentioning one of the first such scientists to take up the new technique was Linus Pauling and whose definition of structural chemistry was clearly indicated by the contents of his book *The Nature of the Chemical Bond* (*18*), i.e. which encompassed a wide vista of chemistry and molecular biology. Not surprisingly, WHB, WLB and Linus Pauling were regarded as being amongst the greatest scientists of the twentieth century. The competitions between WLB, and his research collaborators, with Linus Pauling over the determination of the structures of the silicates, the polypeptide alpha helix and the DNA 3-D structure are also a fascinating but quite another story. The photos of WHB and WLB are shown in Figure 2.5, taken from the Nobel website pages highlighting their Prize; they are portrayed nicely there where one can readily imagine this 'father and son team'.

Acknowledgements

The author is grateful to Prof. Dr Carl Schwalbe who invited him to write a short piece for the newsletter of the British Crystallographic Association *'Crystallography News'*, from which this article is in part derived.

Note

1. A referee remarked: 'In Wikipedia, we learn that Rutherford's first name was meant to be Ernest, but it was mis-spelled as Earnest in the original documentary archives held in New Zealand'.

Notes on the contributor

Professor John R. Helliwell, BA (Physics, York), DPhil (Molecular Biophysics, Oxford), DSc (Physics, York), FInstP, FRSC and FSocBiol. Since 1989, he has been Professor of Structural Chemistry at the University of Manchester becoming Emeritus Professor in August 2012 and self declared as 'semi-retired'. He also worked at the United Kingdom's Synchrotron Radiation Source located at Daresbury Laboratory from 1979 to 1993 and 2003 to 2006, whilst a Joint Appointee with the Universities of Keele, York and Manchester, and also full time as a scientific civil servant (1983–1985) and as CCLRC's Director of SR Science (2002). As an example of his interests in the history of crystallography, he presented the University of Manchester 150th Anniversary 'W L Bragg Lecture' at the Schuster Laboratory in 2001 (*19,20*). The picture of the author was taken during a 'Coast to Coast (Arnside to Whitby)' cycling holiday with Dr Madeleine Helliwell in September 2012. On this holiday, on the final Whitby to Scarborough portion, they alighted on the delightful Cloughton Station Tea Room and Gardens! The photo of John R. Helliwell is taken there. A short history of the station, derived from the Tea Room publicity leaflet gives us a picture of the Bragg family arriving at Cloughton for their summer holiday in 1912, although it is only an assumption that they travelled by train. 'Cloughton Station was built in 1885 and was one of the busiest on the line, having a cattle dock, goods shed, passing line and coal weighbridge. It won many prizes in the annual Best Kept Station competition between 1932 and 1964. Around 21 miles in length, the Scarborough and Whitby Railway opened on 16th July 1885, taking travellers through picturesque coastal and moorland scenery until its closure in 1965'. Currently, the station is run as a guest house, including a converted suite of railway carriages, as well as a Tea Room; see www.cloughtonstation.co.uk.

References

(1) Bragg, W.L. The Structure of Some Crystals as Indicated by their Diffraction of X-rays. *Proc. R. Soc. London, Ser. A* **1913,** *89,* 248–277.
(2) Bragg, W.H. The X-ray Spectrometer. *Nature* **1914,** *94,* 199–200.
(3) Caroe, G.M. *William Henry Bragg 1862–1942 Man and Scientist;* Cambridge University: Cambridge, UK, 1978.
(4) Hunter, G. *Light is a Messenger: The Life and Science of William Lawrence Bragg;* Oxford University: Oxford, UK, 2004.

(5) Jenkin, J. *William and Lawrence Bragg, Father and Son: The Most Extraordinary Collaboration in Science;* Oxford University: Oxford, UK, 2008.

(6) Bragg, W.L. *The Development of X-ray Analysis;* Dover: Mineola, NY, 1975.

(7) Bragg, W.H.; Bragg, W.L. The Reflection of X-rays by Crystals. *Proc. R. Soc. London, Ser. A* **1913,** *88,* 428–438.

(8) Bragg, W.H. The Reflection of X-rays by Crystals. (II.). *Proc. R. Soc. London, Ser. A* **1913,** *89,* 246–248.

(9) Bragg, W.H.; Bragg, W.L. The Structure of the Diamond. *Proc. R. Soc. London, Ser. A* **1913,** *89,* 277–291.

(10) Bragg, W.L. The Analysis of Crystals by the X-ray Spectrometer. *Proc. R. Soc. London, Ser A* **1914,** *89,* 468–489.

(11) Bragg, W.L. X-ray Crystallography. *Sci. Am.* **1968,** *219* (1), 58–70.

(12) Bragg, W.H. The X-ray Spectra Given by Crystals of Sulphur and Quartz. *Proc. R. Soc. London, Ser. A, Containing Papers of a Mathematical and Physical Character* **1914,** *89* (614), 575–580.

(13) Phillips, D.C. W.L Bragg 1890–1971. *Biog. Mem. Fell. R. Soc.* **1979,** *25,* 75–143.

(14) Andrade, E.N.; Da, C.; Lonsdale, K. William Henry Bragg 1862–1942. *Obit. Not. Fell. R. Soc.* **1943,** *4* (12), 276–300, DOI: 10.1098/rsbm.1943.0003.

(15) Bragg, W.H. *The Universe of Light;* G Bell and Sons: London, 1943.

(16) Gatti, C.; Macchi, P., Eds.; *Modern Charge-Density Analysis 2012,* XXIII; Springer: Heidelberg, New York, 2012; p 783.

(17) Cruickshank, D.W.J.; Helliwell, J.R.; Johnson, L.N.; Eds.; *Time-Resolved Macromolecular Crystallography: Proceedings of a Royal Society Discussion Meeting;* Oxford University: Oxford, 1992.

(18) Pauling, L. *The Nature of the Chemical Bond,* 1st ed.; Cornell University: Ithaca, NY, 1939.

(19) Helliwell, J.R. X-ray Crystal Structure Analysis in Manchester from W.L. Bragg to the Present Day. *Z. Kristallogr.* **2002,** *217,* 385–389.

(20) Helliwell, J.R. Lecture demonstrations in a Public Lecture on X-ray Crystal Structure Analysis: From W.L. Bragg to the Present Day. *J. Appl. Crystallogr.* **2009,** *42,* 365. DOI: 10.1107/S0021889809002775. http://www.iucr.org/education/teaching-resources/bragg-lecture-2001.

3

Honouring the two Braggs: The first X-ray crystal structure and the first X-ray spectrometer*

John R. Helliwell†

School of Chemistry, University of Manchester, Manchester M13 9PL, UK

(Received 15 March 2013; final version received 16 April 2013)

In the Centennial celebrations of the birth of X-ray crystal structure analysis, a key feature is to mark the articles which are the first crystal structure analysis studies. This minireview describes the historical development and quotes key statements of W.L. Bragg (WLB) as well as W.H. Bragg (WHB) and the perspectives offered by key players of the time period. The first crystal *layout,* as stated by WLB, is the face-centred cubic arrangement evident in the Laue Laboratory diffraction photographs recorded from a crystal (of zinc blende) and provided to WHB. The first crystal *structure,* as stated by WLB, and explicitly remarked upon by P.P. Ewald, as well as WLB's official biographer, D.C. Phillips, is sodium chloride and which was published in June 1913. The use of the X-ray spectrometer of WHB, and the measurements by WHB, at Leeds University, with this device are acknowledged by WLB in his article. This 1913 article also contains numerous raw diffraction data in the form of 'Laue photographs' measured by WLB of NaCl, and most importantly of KCl, in Cambridge. WLB seemed to anticipate the use of these two isomorphous and closely related alkali halide crystal structures in his article of 1912. The X-ray spectrometer as the forerunner of all X-ray diffractometer designs is also a remarkable initiative of WHB.

Keywords: first X-ray crystal layout; face-centred cubic layout; first X-ray crystal structure; sodium chloride; Laue diffraction photographs; the X-ray spectrometer

Contents

From Crystallography Reviews, Vol. 19, No. 3, May 2013, 108–116.

† Email: john.helliwell@manchester.ac.uk

3.1 Introduction

In the Centennial celebrations of the birth of X-ray crystal structure analysis, a key feature is to mark the article which is the first crystal structure analysis. This mini review describes the historical development and quotes key statements of WLB and the perspectives offered by key players of the time period.

3.2 The words of WLB

An excerpt from WLB's first article in 1912 [1]:

It is only the third point system, the element of whose pattern has a molecule at each corner and one at the centre of each cube face, which will lend itself to the system of planes found to represent spots in the photograph (recorded by Messrs Friedrich and Knipping). This last system, seeing that it forms an arrangement of the closest possible packing, is according to the results of Pope and Barlow the most probable one for the cubic form of zinc sulphide.

Which of these factors it is that decides the form of the interference pattern might be found by experiments with crystals in which the point system formed by the centres of all the atoms differs from that formed by the centres of identical atoms.

In conclusion, I wish to thank Professor Pope for his kind help and advice on the subject of crystal structure.

These concluding words, 'Which of these factors it is that decides the form of the interference pattern', from WLB presciently explained the experiments that would lead to the first crystal structure. From WLB's words quoted by Ewald in [2]:

But let us hear in W.L. Bragg's own words what the exciting sequence of events was after Laue's paper had reached W.H. Bragg in (the) form of an offprint. He tells the story in an address given in 1942 in Cambridge at the first conference on X-ray analysis in industry (held under the auspices of the Institute of Physics), which was published in Science in Britain.

In order to examine the reflected X-ray beam (from a crystal face) more thoroughly, my father (William Henry Bragg (WHB)) built the X-ray spectrometer. The X-ray spectrometer opened up a new world. By using measurements made with the X-ray spectrometer, many of them due to my father, I was able to solve the structures of fluorspar, cuprite, zinc blende, iron pyrites, sodium nitrate and the calcite group of minerals. I had already solved KCl and NaCl, and my father had analysed diamond. Between them, these crystals illustrated most of the fundamental principles of the X-ray analysis of atomic patterns. These results were produced in a year of concentrated work, for the war in 1914 put an end to research. I have gone into these early experiments in some detail because it is a story which I alone can tell, and which I wish to put on record.

These reminiscences can be readily supplemented with the words of WLB in his book completed just two weeks before his death on 1 July 1971. Thus, from WLB (1975), *The Development of X-Ray Analysis*, p. 25 [3]

I found that, although the range of wavelengths represented by the spots did not make sense if one assumed ZnS to be based on a simple cubic lattice everything fell into place if one assumed the basic lattice to be face-centred cubic. ... These results showed, not only that Laue's pictures were made by a continuous range of X-ray wavelengths, a kind of 'white' radiation, but also that X-ray diffraction *could be used to get information about the nature of the crystal pattern.*

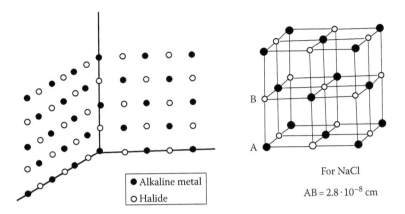

Figure 3.1 The first X-ray crystal structure, NaCl (see [4]). Note: The WLB 1913 paper and his 1975 book have the identical figure, albeit with the left- and right-hand sides switched around, but with the two different figure numbers. Figure 3.1 shows the figure and numbering, from the 1913 article.

The next text page 27 [3] begins with the heading which is Section 5 in Chapter 2 of ref 3 in turn headed 'The Start of X-ray Analysis' and then quotes the last paragraph of this section on page 30. Page 30 [3]:

The First Complete Analyses: The Alkali Halides

It was on this rather indirect and slender evidence that I assigned the structure of Fig 12 (see Figure 3.1) to the alkaline halides in a paper read to The Royal Society in June 1913 [4]; fortunately further investigation established its correctness! These were the first crystals to be analysed by X-rays. As the structure was now established, it was possible to calculate dimensions from the crystal density and the mass of the NaCl molecule. Half a molecule is associated with each small cube of side a = AB in Fig 12 (Figure 3.1) so

$$\frac{1}{2}Mm = \rho a^3,$$

where M is the molecular weight, m is the mass of the hydrogen atom, and ρ is the density of the crystal. This gave a value for a of 2.8×10^{-8} cm and so established a scale for the measurement of all X-ray wavelengths and crystal spacings.

There is now a chapter devoted to the first crystal structure analysis:

Page 53 [3]:

Chapter 5 'The First Analysis of Crystal Structure'

The Method of Analysis

Although the NaCl structure was deduced from Laue photographs, the first results with the X-ray spectrometer showed at once how far more powerful it was as an analytical tool. When I started work in the Leeds laboratory in the summer of 1913, my father was still mainly interested in exploring the X-ray spectra. It fell to me to use the spectrometer [5] for determinations of crystalline arrangement and a number of inorganic structures were discovered. We wrote a joint paper on diamond [6] and the other structures were described in a paper in The Royal Society Proceedings in 1913 which may be said to represent the start of X-ray crystallography.

It was very fortunate for me that I was able to work in my father's laboratory. Young research students nowadays can have little conception of the primitive conditions in a research laboratory some sixty years ago. (However) In my father's laboratory ... at Leeds there was a good workshop with an excellent mechanic in charge to carry out his ideas. It was the privilege of working with really effective apparatus which made it possible for me to start my research career by working out a number of crystal structures. ... The analysis depended on comparing the strength of the various orders of reflection. When the planes are identical and evenly spaced the orders fell off regularly... A marked departure from this regular diminution indicated that the planes were not simple. ... The (crystal structures analysed) included fluorspar, zinc blende, pyrites and calcite (in various forms).

3.3 The words of WHB

WHB's own words, extracted from the article [7] of his son's role (see Figure 3.2):

From the work now described by W L Bragg it appears that the reflection phenomena lead to a more definite knowledge of crystal structure, and we may now complete various quantitative determinations.... (namely) the (X-ray) wavelengths of other homogeneous rays can then be found easily as soon as their angles of reflection are known (from an NaCl or other single crystal).

[The unit cell parameter for the cubic NaCl having been established ingeniously from the mass of a crystal, its volume and the atomic weights of sodium and chlorine, as described above.]

The X-ray spectrometer was also noted by WHB to be under similar development and use at Liverpool University by Barkla and at Manchester University by Moseley and Darwin; for a summary description see [8].

3.4 The words of P.P. Ewald

Page 65 of 'Fifty Years of X-ray Diffraction [2]':

Although this early paper (WLB 1912 [1]) does not yet contain a full structure determination, it comes very close to one, in the case of such a simple compound as ZnS.

Page 69:

The great break-through to actual crystal structure determination and to the absolute measurement of X-ray wavelengths occurred in W L Bragg's (NaCl) paper [4].

Page 71:

In the series of fundamental papers published by both Braggs in 1913 and 1914 this paper by W L Bragg unquestionably brings the greatest single advance ... it made all future structure determinations very much easier by providing an absolute wave-length scale ... It would, however, be an invidious undertaking to single out any one of the early papers as the most important one, so closely were they all interlinked and so rapid was the progress at the time of their writing which formed a background for their formulation.

Page 72:
The joint paper The Structure of Diamond [6]:

was the first example of a structure in which the effective scattering centres did not coincide with the points of a simple (Bravais type) lattice. The determination of this structure was acclaimed as a great triumph of the new methods. Whereas in the structures of rock salt, zinc blende and fluorite, the

The Reflection of X-rays by Crystals. (II.)

By W. H. BRAGG, M.A., F.R.S., Cavendish Professor of Physics in the
University of Leeds.

(Received June 21,—Read June 26, 1913.)

This note is a supplement to a paper on the reflection of X-rays by crystals which has been recently communicated to the Royal Society.* It is there shown that the wave-length of a homogeneous beam of X-rays can be found accurately in terms of the spacing of the elements of a crystal. There has been some doubt as to the actual arrangement of the atoms in the crystal and in consequence it was not possible in the paper quoted to draw any final conclusions as to wave-length values. From the work now described by W. L. Bragg it appears that the reflection phenomena lead to a more definite knowledge of crystal structure, and we may now complete various quantitative determinations.

The elementary volume in rock-salt is a cube with 1 atom of sodium at each of four corners and 1 atom of chlorine at each of the other four. In other words the number of elementary volumes in any space of measurable dimensions is equal to the number of atoms in that space.

The number of molecules in 1 c.c. of NaCl is

$$2 \cdot 15/58 \cdot 5 \times 1 \cdot 64 \times 10^{-24} = 2 \cdot 24 \times 10^{22}$$

(The weight of the H atom is taken to be $1 \cdot 64 \times 10^{-24}$.)

The number of atoms is twice as great and the elementary cube volume is therefore $1/4 \cdot 48 \times 10^{22} = 2 \cdot 23 \times 10^{-23}$. The edge of the cube is $2 \cdot 81 \times 10^{-8}$; this is the distance between consecutive reflecting planes parallel to (100).

The principal bundle of homogeneous X-rays from a platinum anticathode is stated in the paper quoted to be reflected at the (100) face of rock-salt at a glancing angle of $11 \cdot 55°$. Recent observations with better apparatus show that this bundle is really double, consisting of two separate sets whose wave-lengths differ from each other by a little less than 2 per cent. of either; they also show that the first estimate was a little too high. For the purpose of the present argument it is sufficiently accurate to ignore the division and assume the angle to be $11 \cdot 3°$. This gives a wave-length

$$(2 \, d \sin \theta) = 2 \times 2 \cdot 81 \times 10^{-8} \times 0 \cdot 196 = 1 \cdot 10 \times 10^{-8}$$

The wave-lengths of other homogeneous rays can then be found easily as soon as their angles of reflection are known.

* W. H. Bragg and W. L. Bragg, these 'Proceedings,' A, vol. 88, p. 428.

Figure 3.2 From [7]. Further text extracts of the various papers from WLB and/or WHB are in the centennial celebration article [8].

absence of molecules in the accepted sense created an element of bewilderment, the beautiful confirmation of the tetravalency of carbon on purely optical principles made this structure and the method by which it was obtained immediately acceptable to physicists and chemists alike.

Page 73:

The paper The Analysis of Crystals with the X-ray Spectrometer [9] shows remarkable progress in a number of ways ... (this included the fact that) it is clearly recognized that for a complete structure analysis the intensities of the reflections have to be known and evaluated.

3.5 The words of D.C. Phillips

David Phillips, who knew WLB directly, e.g. at The Royal Institution, wrote the Biographical Memoir of WLB in The Royal Society series [10]. On pages 88–89 referring to WLB (1912) he states

The critical test was to see whether these ideas (of WLB) explained the observations from (the) ZnS (Laue diffraction photographs), including the absence of some spots predicted by Laue's analysis. Here (W L) Bragg inverted the argument and used the fact that the X-ray pulses can be regarded to be a 'white light' spectrum extending over a characteristic range of wavelengths and with maximum energy at certain wavelengths. The intensities of the Laue spots ought, therefore, to fall in a regular series depending upon which part of the spectrum was responsible for each of them. Examinations showed that this did not work.... (W L) Bragg tried to explain the ZnS pattern (of diffraction spots) on the assumption that the structure is face-centred cubic and everything fell into place. Thus he showed that the Laue pictures were made by a continuous range of X-ray wavelengths ... and that X-ray diffraction could be used to get information about the crystal structure. This was the start of the X-ray analysis of crystals ... The next papers were published at about the same time (June 1913). In the first of them WHB derived the wavelengths of various radiations and correlated them with Barkla's characteristic radiations, making use of the structure of rock salt which had been worked out by his son, but not yet published. This paper was immediately followed by Bragg's detailed account [4] of NaCl and related structures described by Ewald [2] as 'the great breakthrough to actual crystal structure determination and to the absolute measurement of X-ray wavelengths'. The analysis depended mainly on Laue photographs taken in Cambridge, supported by some measurements with the (WHB) spectrometer.

3.6 Other perspectives

The Royal Society Obituary Notice for WHB was written by Andrade and Lonsdale [11]. It contains a synoptic paragraph of their achievements (Figure 3.3). The emphasis given in [11] between WLB and WHB, son and father, in that synopsis seems not to be correct not least with respect to WHB's own words of his son's work (see above) from [7].

This perhaps explains the modern comment of the type:

11 November 2012 marks the centenary of the reading of the paper by William Lawrence Bragg (WLB) to the Cambridge Philosophical Society outlining the foundations of X-ray crystallography. It included the derivation of the first correct atomic structure of a crystal, namely that of zinc blende, based on the X-ray diffraction pattern recorded by Friedrich, Knipping and Laue in the spring of 1912.

The work of Bragg and his son Lawrence in the two years 1913, 1914 founded a new branch of science of the greatest importance and significance, the analysis of crystal structure by means of X-rays. If the fundamental discovery of the wave aspect of X-rays, as evidenced by their diffraction in crystals, was due to Laue and his collaborators, it is equally true that the use of X-rays as an instrument for the systematic revelation of the way in which crystals are built was entirely due to the Braggs. This was recognized by the award of the Nobel prize for Physics in 1915 to them jointly 'pour leurs recherches sur les structures des cristaux au moyen des rayons de Roentgen', and a further formal acknowledgment was the appearance in Leipzig, in 1928, of a collected reprint, in German translation, of the early papers, under the title *Die Reflexion von Röntgenstrahlen an Kristallen: grundlegende Untersuchungen in den Jahren* 1913 *und* 1914 *von W. H. Bragg und W. L. Bragg.*

Figure 3.3 Text extract from [11] by Andrade and Lonsdale.

This comment can be read recently: in Acta Cryst Section A and in at least two newsletters of national crystallographic societies or listened to on the radio by scientific commentators.

Perhaps most remarkably, the following appeared in the paperwork for the 2011 IUCr Madrid Congress of the Crystallography General Assembly papers opening the item on the International Year of Crystallography:

In 1912 Max Laue showed that X-rays were diffracted by crystals and W.L. (Lawrence) Bragg presented a paper to the Cambridge Philosophical Society both presenting Bragg's Law and the correct structure of zinc blende, which he derived from the X-ray diffraction data obtained from Laue. W.H. and W.L. Bragg rapidly carried out a number of key diffraction experiments of their own that led to the determination of crystal structures that were published in 1913. These ground breaking experiments mark the birth of modern crystallography. The International Union of Crystallography (IUCr) is marking the centennial of these events by declaring 2013 the International Year of Crystallography (IYCr2013).

If the substance of the modern comments highlighted just above were to restrict themselves to *the layout of a crystal*, it would be factually correct, namely face-centred cubic in the case of WLB's 1912 analysis [1]. Those modern recent comments however, arguably much more important, also seem to basically ignore the importance of the X-ray spectrometer and the monochromatic X-ray measurements from the crystals that it empowers. The Laue method of white X-rays, as WLB pointedly remarks [1], is very sensitive to the crystal orientation, whereby the intensities of individual reflections are altered. Of course, the systematically absent reflections have zero intensity whatever be the crystal orientation and which is why the Laue method was adequate for WLB to use it to determine the face-centred cubic lattice layout from the diffraction photographs [1] provided by Laue to WHB. Indeed the wavelength normalization of the Laue photographic intensities has been a key step of the modern synchrotron Laue method of complete quantitative crystal *structure* analyses [12]. Again, as WLB remarks, the move towards more complicated crystal structures, like diamond, in 1913 into 1914 was firmly implanted within the use by WLB, along with his father, of the X-ray spectrometer.

Further extensive documentation of the achievements of WLB is available in the volume by Thomas and Phillips [13] and the Special Issue of Acta Crystallographica Section A: Foundations of Crystallography January 2013 [14]. A tribute to the work in X-ray analysis of WHB is the scientific summary written by North [15]. A further, recent, tribute to the work of WLB is given by Thomas [16] and from which I quote

1913 marks the year when X-ray crystallography, through the determination of the structures of sodium chloride, potassium chloride, potassium bromide and potassium iodide and of diamond first made its striking impact.

3.7 Conclusions

This Historical Note honours the first crystal structure, that of WLB's NaCl, published in June 1913 [4]. The earlier, also truly remarkable, analysis by WLB of the face-centred cubic layout of a crystal from the pattern of systematic absences in the Laue diffraction photographs of ZnS [1] included WLB's anticipation in his concluding paragraph of what studies and experiments he would have to make to determine the first crystal structure, i.e. with the placement of the Na and Cl atoms [4]. The prompt subsequent development of the X-ray spectrometer apparatus by WHB in Leeds [5], and its use by his son and by himself, led to the more quantitative analysis of the monochromatic X-ray measurements from single crystals. The first X-ray crystal structure, that of NaCl, described vividly in WLB's own words [2,3] is a truly absorbing, wondrous, iconic story in the History of Crystallography and indeed of all of science. The critical nature of the contribution of a new piece of apparatus, the X-ray spectrometer [5], and the measurements it empowered in these studies is also a wondrous development.

Notes on the contributor

John R. Helliwell trained as a physicist (York University, 1st class Hons degree) and then a protein crystallographer (DPhil., Oxon). He spent the first 20 years of his research career, from 1979, walking in the footsteps of William Henry Bragg with the design and build of the Daresbury Synchrotron Radiation Source Station 7.2 tuneable synchrotron X-ray spectrometer for protein crystallography. This synchrotron radiation X-ray spectrometer was followed by SRS wiggler Station 9.6, also tuneable, but extended to shorter X-ray wavelengths, then SRS wiggler Station 9.5 which was rapidly tuneable and finally the rapidly tuneable *and* high X-ray intensity of the instrument ESRF BM14 (this latter project led by Andrew Thompson). In the last 20 years, JRH switched emphasis and followed in the footsteps of William Lawrence Bragg, based at the University of Manchester from 1989 as Professor of Structural Chemistry, and where he has undertaken a wide variety of crystal structure analyses of proteins and nucleic acids, using X-rays and more recently with neutrons. He is now an Emeritus Professor at the University of Manchester.

References

[1] Bragg WL. The diffraction of short electromagnetic waves by a crystal. Proc Camb Phil Soc. 1912;XVII (1):43–57 [Communicated by Sir J.J. Thomson].
[2] Ewald PP. The immediate sequels to Laue's discovery, Chapter 5. In: Ewald PP, editor. Fifty years of X-ray diffraction. Utrecht: Published for the International Union of Crystallography by N V A Oosthoek; 1962.
[3] Bragg WL. The development of X-ray analysis. New York: Dover Publications; 1975.
[4] Bragg WL. The structure of some crystals as indicated by their diffraction of X-rays. Proc R Soc Lond A. 1913;89:248–277.
[5] Bragg WH. The X-ray spectrometer. Nature. 1914;94:199–200.
[6] Bragg WH, Bragg WL. The structure of diamond. Proc R Soc Lond A. 1913;89:277–291.
[7] Bragg WH. The reflection of X-rays by crystals. (II.). Proc R Soc Lond A. 1913;89:246–248.
[8] Helliwell JR. The centennial of the first X-ray crystal structures. Crystallogr Rev. 2012;18(4):280–297.
[9] Bragg WL. The analysis of crystals with the X-ray spectrometer. Proc R Soc A. 1914;89:468.
[10] Phillips DC. W L Bragg 1890–1971. Biograph Memoirs Fellows R Soc. 1979;25:75–143.
[11] Andrade ENdaC, Lonsdale K. William Henry Bragg. 1862–1942. Obit Not Fell R Soc. 1943;4(12): 276–300.

[12] Helliwell JR, Habash J, Cruickshank DWJ, Harding MM, Greenhough TJ, Campbell JW, Clifton IJ, Elder M, Machin PA, Papiz MZ, Zurek S. The recording and analysis of Laue diffraction photographs. J Appl Crystallogr. 1989;22:483–497.

[13] Thomas JM, Phillips D, editors. Selections and reflections: the legacy of Sir Lawrence Bragg. Middlesex: Science Reviews Ltd.; 1990.

[14] Wilkins SW. Editor Bragg centennial a special issue of. Acta Crystallogr Sect A. 2013;69:1–62.

[15] North ACT. William Henry Bragg: grandfather of X-ray analysis. Univ Leeds Rev. 1976;19:125–148.

[16] Thomas JM. William Lawrence Bragg: the pioneer of X-ray crystallography and his pervasive influence. Angew Chem Int Ed. 2012;51:12946–12958.

Section III

Aspects of crystallography research

4

FULL REVIEW
The evolution of synchrotron radiation and the growth of its importance in crystallography*†

John R. Helliwell‡

School of Chemistry, University of Manchester, Manchester, UK

(Received 19 August 2011; final version received 9 October 2011)

The author's 2011 British Crystallographic Association Lonsdale Lecture included a tribute to Kathleen Lonsdale followed by detailed perspectives relevant to the title, with reference to the Synchrotron Radiation Source (SRS) and European Synchrotron Radiation Facility (ESRF). Detector initiatives have also been very important as have sample freezing cryomethods. The use of on-resonance anomalous scattering, smaller crystals, ultra-high resolution as well as the ability to handle large unit cells and the start of time-resolved structural studies have allowed a major expansion of capabilities. The reintroduction of the Laue method became a significant node point for separate development, and has also found wide application with neutron sources in biological and chemical crystallography. The UK's SRS has now been superseded by Diamond, a new synchrotron radiation source with outstanding capabilities. In Hamburg we now have access to the new ultra-low emittance PETRA III, the ultimate storage ring in effect. The ESRF Upgrade is also recently funded and takes us to sub-micrometre and even nanometre-sized X-ray beams. The very new fourth generation of the X-ray laser gives unprecedented brilliance for working with nanocrystals, and perhaps even smaller samples, such as the single molecule, with coherent X-rays, and at femtosecond time resolution.

Keywords: Kathleen Lonsdale; synchrotron radiation; high brightness; neutrons

Contents

---

* From *Crystallography Reviews* Vol. 18, No. 1, January 2012, 33–93.

† The British Crystallographic Association Lonsdale Lecture and Teaching Plenary 2011.

‡ Email: john.helliwell@manchester.ac.uk

48   *J. R. Helliwell*

4.6   The importance of the second- and third-generation SR sources in bringing about a revolution in crystallography 68
  4.6.1   SRS Station 7.2: the first protein crystallography instrument on a dedicated SR X-ray source 68
  4.6.2   SRS Station 9.6: the first super conducting wiggler (SCW) protein crystallography instrument and electronic area detector initiatives 72
  4.6.3   The SRS High Brightness Lattice and SRS 9.5 for rapidly tuneable protein crystallography and point-focussed Laue crystallography 74
  4.6.4   The ESRP and the ESRF planning 76
  4.6.5   ESRF BM14: the first macromolecular crystallography instrument on a third-generation SR source 80
  4.6.6   SR Laue crystallography at SRS and ESRF 81
  4.6.7   Synergies of Synchrotron and Neutron Laue macromolecular crystallography: initiatives at the Institut Laue Langevin 83

4.7   The impacts of the SRS and ESRF in macromolecular crystallography 87

4.8   Other relevant topics 88

4.9   Research directions for the future 89

Acknowledgements 90

Notes on the contributor 91

Summary of abbreviations used 91

References 92

4.1 Introduction

This review article is based on the Lonsdale Lecture and Teaching Plenary that was presented by the author under the auspices of the Biological Structures Group of the British Crystallographic Association (BCA) at its Annual Conference held at Keele University in April 2011. This event was chaired by the BCA President, Professor Elspeth Garman. This article is divided into two parts. Part 1 is a short review of the previous Lonsdale Lecturers and also a glimpse of the tributes made to Kathleen Lonsdale's achievements as well as a short description of her broad interests spanning not only crystallographic science but also her pacifism. Part 2 is my review of synchrotron radiation (SR) and crystallography involving the perspectives that I gained, and their allied topics, at the Synchrotron Radiation Source (SRS) and European Synchrotron Radiation Facility (ESRF), the first dedicated SR X-ray source and the first third-generation low-emittance SR source, respectively.

My first preparation before my Lecture was to read Dorothy Hodgkin's Royal Society Biographical Memoir of Kathleen Lonsdale (1); Dorothy also features in a pivotal position in my account below.

4.2 Previous Lonsdale Lecturers

The Lonsdale Lecture originated as a special Lecture offered by the BCA to the Annual Meeting of the British Association for the Advancement of Science (BAAS). They were announced in *Acta Crystallographica* (2a) as follows:

As a result of a suggestion (by Moreton Moore) from the Bragg Lecture Fund committee, the Kathleen Lonsdale Lectures have been established by the British Crystallographic Association to commemorate her achievements. These lectures are intended to educate the public in the science of crystallography and will be given at the annual meetings of the British Association. The first one will be at 2 pm on 27 August 1987 at the British Association meeting in Belfast, Northern Ireland, and will be open to the public. The lecture will be given by Professor David Blow and the title of the lecture is 'Protein Crystallography Applied to Medicine and Industry'.

From the mid-1990s to late 1990s onwards, the BCA Council decided to host the Lonsdale Lecture within the BCA Annual Conference. In recent years, the Young Crystallographers Group of the BCA was formed, and who started nominating the Lonsdale Lecturer to the Council of the BCA whereby the speaker is 'a well-respected scientist who has a good rapport with students'.

From the BCA website and BCA Newsletters, the past Lonsdale Lecture details are

- 1987 David Blow (Imperial College) – Belfast BAAS meeting (*2a*)
- 1988 Michael Hart (Manchester University) – Oxford BAAS meeting Sept 5th entitled *Synchrotron radiation throws light on a microscopic world*
- 1989 Robert Diamond (Laboratory of Molecular Biology, Cambridge) – Sheffield BAAS meeting 12th September 1989 entitled 'Crystalline Viruses'.
- 1990 Alan MacKay (Birkbeck College) – Swansea BAAS meeting August 22nd entitled *What is a crystal?*
- 1991 Louise Johnson (Oxford University) – Plymouth BAAS meeting August 29th entitled *Designer Drugs*
- 1992 Peter Murray-Rust (Birkbeck College) – Southampton BAAS meeting entitled *Molecules and Disease*
- 1993 Judith Milledge (University College London) – entitled *Diamonds thick and thin* Keele BAAS meeting
- 1997 Andrew Lang (Bristol University) – presented at the Leeds BCA Annual meeting entitled *X-rays and Diamonds*
- 1999 T. Richard Welberry (Australian National University, Canberra, Australia) – presented at the IUCrXVIII held in Glasgow entitled *Diffuse X-ray Scattering*
- 2004 Elspeth Garman (Oxford University) presented at the BCA held in Manchester entitled *Cool crystals: kill or cure?*
- 2006 Mike Glazer (Oxford University) presented at the BCA held in Lancaster entitled *Crystals under the microscope*
- 2007 Bill David (ISIS neutron facility, Rutherford Appleton Laboratory) presented at the BCA held in Canterbury entitled *Combinatorial Studies of Hydrogen Storage Materials.*
- 2009 David Watkin (Oxford University) presented at the BCA held in Loughborough entitled *Chemical crystallography – science, technology or a black art* (*2b*).

4.3 Biographical sketch of Kathleen Lonsdale and her crystallographic science

Dame Kathleen Lonsdale FRS (Figure 4.1) was a key figure in British crystallography and science as indicated by a list of a few examples of her accomplishments: President of the IUCr 1966; a key instigator of International Tables; author of numerous crystallography publications and the first female President of the BAAS. She was also a leading pacifist of her time. She trained as a physicist and became a Professor of Chemistry.

Figure 4.2 shows the front cover of her book 'Crystals and X-rays' (*3*). In the book review by Prof. Martin Buerger (*Department of Geology, Massachusetts Institute of Technology, Cambridge, MA, USA*) (*4*), he wrote:

This little book (Pp. xxviii+ 195.) is divided into seven chapters: I, Historical Introduction. II, Generation and Properties of X-rays. III, The Geometry of Crystals: X-ray Methods of Investigation. IV, Geometrical Structure Determination. V, Determination of Atomic and Electronic Distribution. VI, Extra-structural Studies. VII, The Importance of the Study of Crystals.... The book is based on

Figure 4.1 Dame Kathleen Lonsdale FRS (28 January 1903–1 April 1971); Kathleen Lonsdale at the time of her election to The Royal Society in 1945. With the permission of The Royal Society and Highwire Press © Godfrey Argent Studio.

Figure 4.2 The front cover of (my own copy of) her book, ref. (*3*).

a series of public lectures given by the author at University College, London.... For the scientist not versed in X-ray crystallography, perhaps the most important chapter is the comparatively non-technical 'Historical Introduction'. In this chapter the author not only sketches the history of the development of X-rays and X-ray diffraction, but gives a nicely balanced view of the place of X-rays in modern science. When the evidence is assembled, the debt modern science owes to X-rays and X-ray diffraction is striking indeed. In the reviewer's opinion, this chapter is beautifully done and should be read by all scientists.... The veteran X-ray crystallographer will probably find that a good deal of the book is a discussion of material quite familiar to him. Yet almost every reader will find a smattering of this subject matter which appears in novel and stimulating form.

Figure 4.3 is one of her famous crystallographic analyses, namely that of hexachlorobenzene (*5*), which followed her 1929 crystal structure of hexamethylbenzene (*6*), analyses which showed that the bond distances around the aromatic benzene ring were equal and not alternating short and long distances of single and double bonds. Thus, it is a resonance molecular structure. The advantage of the hexachlorobenzene structure is in showing the larger number of (relative scale) contours for the terminal chlorine atoms than the parent carbon atoms. Kathleen Lonsdale is careful to explain in ref. (*6*) that *'It must be remembered, however, that these values are measured in arbitrary units and are subject to the addition of an unknown constant. Any attempt at electron counting, for example, is out of the question'.*

I now offer examples that illustrate the breadth of her interests and work.

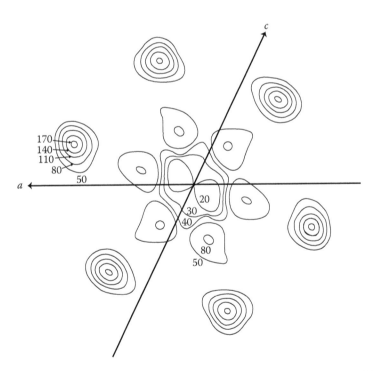

Figure 4.3 Kathleen Lonsdale published in 1931 an X-ray analysis of the structure of hexachlorobenzene by the Fourier method (*5*) illustrated here with her Figure 7 using the formula: $S(x, z) = \sum \sum F(h0l)$ cos $2\pi(hx/a + lz/c)$. The unit cell is simple monoclinic prismatic, with $a = 8.07$, $b = 3.84$, $c = 16.61$ Å, $\beta = 116°52'$ and it contains two molecules of C_6Cl_6. This figure is reproduced with the permission of The Royal Society and HighWire Press and is Figure 7 in ref. (*5*).

Her obituary in *The Times* (*7*), in noting that Kathleen Lonsdale was the 'First woman President of the British Association for the Advancement of Science', stated that '(she) aimed darts at a variety of targets: the sale of arms, the narrowness of many scientists and (the) responsibility of scientists as a whole for the use made of their discoveries.' In the same role she delivered a lecture at the Leeds meeting of the BAAS on the 4 September 1967 on *Physics and Ageing.*

As a pacifist, in her letter to the Minister of Labour and National Services, Whitehall SW1 of 29 May 1942 she wrote (*8*):

Sir,

…

I am absolutely opposed, however, to the principle of <u>universal</u> compulsory registration and conscrip-tion <u>for war purposes</u>. As a member of the Society of Friends (Quakers) I believe war to be the wrong method of resisting aggression or any other form of evil … In refusing, therefore, to comply with the regulation or to take advantage of possible exemption on conscientious or other grounds, and in being prepared to take the consequences, I am acting in accordance with what I believe to be my highest duty. I rather wish I did not. Yours sincerely,

Kathleen Lonsdale (D.Sc London).

She served 1 month in Holloway Prison in 1943 as a result.

She expressed her pacifist views in detail in her book (*9*): 'Is Peace Possible?', which I came across in writing up my Lecture in this article. I bought a copy and it is a personally signed one. This signature, like the two letters that I highlight below, gives me a real sense of her presence. The front cover of the book has the headline: *A Quaker scientist discusses problems of peace, freedom, and justice in an era of expanding world population and technical development.* She researched her topics carefully; she had clearly studied and quotes from the UN Charter as well, for example, of giving a detailed summary of the stages and landmarks in the Israeli–Palestinian conflict, as well as the role of Britain and other countries in that conflict (Chapter 9). The book also contains her strongly held religious beliefs. She sums up her arguments and logic made in the book as follows (p. 127):

… a life of non-violence is essentially one of deep spiritual out-reach to the good in other men and of belief that, even if there is no response, even if we appear to fail, goodness will in the end prevail. *Yea, though I walk through the valley of the shadow of death I shall fear no evil, for Thou art with me.*

In covering her science, I must mention her strong interest in determining the mobility of atoms from crystal structure analyses, and thereby inferring details of reactions in crystals and the crys-talline state. In her article with Judith Milledge '*Analysis of thermal vibrations in crystals: a warn-ing*' (*10*), the authors are concerned with assumptions regarding thermal ellipsoid determinations from crystal structures and conclude that 'it is *not* safe to rely on one determination of a set of b_{ij} deduced from experimental data at one temperature …Finally we would beg crystallographers not to mislead interested research workers in other disciplines …unless and until they repeated their measurements and refinements more than once, independently, and can prove that a claim to such accuracy (4 or 5 apparently significant figures) is justified'. In their acknowledgements, they refer to their 'indebtedness to the referee whose sharp criticisms made us rewrite this paper and, we hope, improve it'. Many years later, the IUCr Newsletter published a piece from Prof. Durward Cruickshank entitled 'tilting at Windmills (*11*) in which Durward revealed that he was the referee and furthermore, a sketch drawn by Kathleen Lonsdale was reproduced; evidently, the matter ended in good humour. Furthermore, I would surmise, the general impression left with Kathleen Lonsdale of Durward Cruickshank's abilities and conscientiousness led to the letters to him when she was the President of the International Union of Crystallography dated 14 June 1966, and subsequently (Figure 4.4a and b). Durward also wrote of Kathleen Lonsdale in that IUCr Newsletter piece (2005) (*11*):

> Kathleen Lonsdale (1903–1971) began her long career in X-ray crystallography under W.H. Bragg in 1922. In 1929 she determined the structure of hexamethylbenzene, the first aromatic compound to be defined by X-ray diffraction, and so proved the planarity of the benzene nucleus. She was a pioneer of space group theory in relation to the structure analysis of crystals and was an indefatigable Editor of International Tables for many years. Among other fields she was an expert on thermal diffuse scattering, divergent beam X-ray photography and synthetic diamonds. At the Moscow Congress in 1966 she became President of the IUCr. Dorothy Hodgkin wrote in her obituary "She was a tough President, who kept her committees working for long hours. Professor Belov, who succeeded her as President, was heard to remark as she dragged him away from a pleasant party at the British Embassy to yet another committee meeting: 'Kathleen you are a martinet'".

W.L. Bragg wrote one of his last communications before he died regarding the passing of Kathleen Lonsdale, and so, I can quote an extract written by him published in *Acta Cryst.* (*12*):

> Others have written about Dame Kathleen Lonsdale's scientific achievements; I wish to add a tribute to her as a person. She was one of the most thorough, high-principled, and courageous people I have known. We who work in the field of X-ray analysis cannot be too grateful for all she did to help us. Her early collaboration with Astbury in preparing an exhaustive survey of space groups was typical. The set of formulae for structure analysis, and the work on *International Tables* for the Union, were models of accuracy and ordered arrangement. No trouble was too great for her, and it was all done disinterestedly for the general good, much of it behind the scenes and in helping others …. Her pluck and determination to be of service to others knew no bounds.

In closing this section, I wish to quote extracts from one of her devoted colleagues, Dr Judith Milledge, both from her email to me during the preparation of this article and her obituary of Kathleen Lonsdale.

From Judith Milledge's email to me (*13*):

> You must obviously choose which of her achievements you mention, but being one of the first female FRS' [Fellows of the Royal Society], the 1st woman professor at University College London, and her contribution to teaching crystallography in the joint MSc with Bernal at Birkbeck, as well as the week-long crystallography practical for 3rd year chemistry students, which was copied by many institutions, might be worth a mention …. My article in 'Out of the shadows' Edited by Byers & Williams (C.U.P. 2006, pp. 191–201 (*14*)) also mentions that her diamagnetic anisotropy measurements provided the first direct experimental confirmation of the existence of molecular orbitals …. You are correct in assuming that our interchange with Durward ('tilting at windmills') ended in good humour.

From Judith Milledge's Obituary of Kathleen Lonsdale, *Acta Cryst.* (*15*):

Firstly on Kathleen Lonsdale's approach: 'Her exceptional intellect and relentless logic drove her towards practical solutions of problems once she had become involved with them, and her almost infinite capacity for hard work ensured that once a course of action had been determined, progress was steady'.

Secondly, as a complement to the above descriptions of what she did, here is a summary in the obituary on what she was not interested in: 'She never took the slightest interest in research on defect controlled crystal growth, refused the offer of good electron diffraction equipment in return for training students to use it, had no interest in the derivation of atomic scattering factors, potential

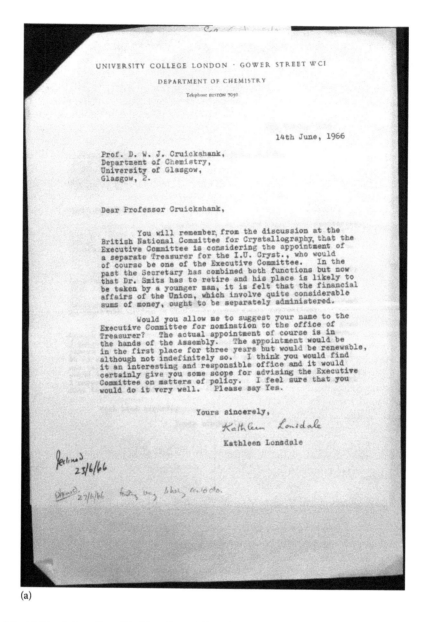

(a)

Figure 4.4 (a, b) Two letters from Kathleen Lonsdale to Durward Cruickshank when she was President of the International Union of Crystallography. Notes: These two letters are in the Durward Cruickshank letters archive donated by Durward to the University of Manchester and now properly listed and curated here at our Library, the John Rylands University of Manchester Library (JRULM). These letters are reproduced here with the permission of the JRULM on the understanding that I have the relevant permissions. Thus, I have obtained the permission of the IUCr, under whose auspices they were sent since Kathleen Lonsdale was writing as President of the IUCr; from Dr Judith Milledge of University College London, as Scientific Executor for Kathleen Lonsdale; and from the family of Durward Cruickshank, namely, his son, John Cruickshank and his daughter Helen Stuckey (née Cruickshank).

I am keen for readers to see Kathleen Lonsdale's nice handwriting but to help any reader with legibility if needs be here is a typed version. *Dear Durward, The Executive Committee seems at present to have no other nomination for Treasurer and they welcome your willingness to serve. They realize that your own circumstances may change. If you feel that this change, if and when it occurs, obliges you to resign, that will be understood and there is a procedure for dealing with that situation. On the other hand, a change of residence or nationality need not oblige you to resign. It would be a matter for consultation within the framework of the statutes and By-laws; and we can meet that situation when it and if it arises! Yours sincerely, Kathleen Lonsdale.* (Continued)

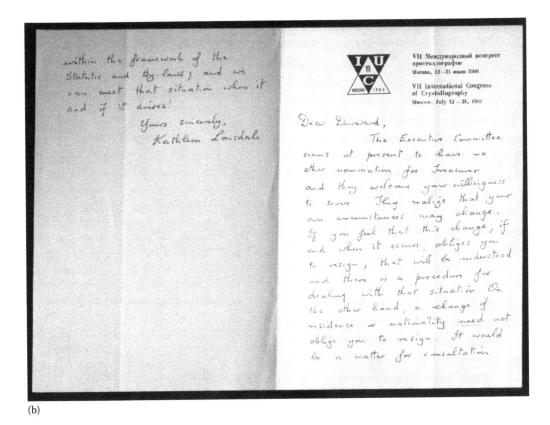

within the framework of the
Statutes and By-laws; and we
can meet that situation when it
and if it arises!
    Yours sincerely,
    Kathleen Lonsdale

VII Международный конгресс
кристаллографов
Москва, 12—21 июля 1966

VII International Congress
of Crystallography
Moscow, July 12—21, 1966

Dear Durward,

The Executive Committee
seems at present to have no
other nomination for Treasurer
and they welcome your willingness
to serve. They realize that your
own circumstances may change.
If you feel that this change, if
and when it occurs, obliges you
to resign, that will be understood
and there is a procedure for
dealing with that situation. On
the other hand, a change of
residence or nationality need not
oblige you to resign. It would
be a matter for consultation

(b)

Figure 4.4 (a, b) *(Continued)*

functions or any of the many types of spectra obtainable from crystals, and made no major contributions to the development of direct methods or other structure-solving techniques.

She remained in essence a crystal physicist, and apart from W.H. Bragg, the most important influence on her scientific outlook was probably Michael Faraday. Working in Faraday's room at the Royal Institution, she read his notebooks and absorbed his approach to experimentation. She often reminded me that Faraday had discovered all the important laws governing electromagnetic phenomena accessible with the apparatus available to him without recourse to anything other than simple arithmetic. She was herself a great experimentalist, and in her later years became less and less attracted to complicated mathematical formulations of problems which she felt could be tackled directly from first principles; 'One of us', she wrote at the end of a paper in 1959 'would like here to acknowledge personal indebtedness to the late Sir William Bragg for a training which emphasized that a simple approach to a difficult problem is not necessarily an inaccurate one'.

[Of course on the matter of the power of arithmetic, mentioned above, versus complicated mathematical formalisms in crystallography here, I would beg to differ with her to emphasize in my view the power of direct methods and maximum likelihood in phasing and refinement with modern computers and software. She might of course have steered me into a discussion on brute force computing, such as the power of Monte Carlo ray tracing in X-ray optics design, and the understanding of their aberrations. Thus, of course, elegance of the mathematics is not always a guarantee of utility, I would agree.]

Thirdly, in summing up, Judith Milledge wrote:

'Those accustomed to equate religious convictions with unworldliness often had a rude awakening when dealing with Kathleen, who could be a very tough negotiator, and had always done her homework on the committee papers. "It's lucky one of us is an idealist" she would remark, while coping expertly with the practical side of some rarefied gathering, because she had an unshakeable conviction that the right course of action in any circumstances was by definition also the most practical one. She also had a great sense of humour and a most perceptive appreciation of the best things in life. She concluded her Inaugural Lecture at University College by explaining that crystallographers were, like Walter de la Mare's Wizards "a flock of crazy prophets who by staring at a crystal can fill it with more wonders than are herrings in the sea". She did her share.'

In closing this part 1 of my article, I wish to conclude with my own tribute to Kathleen Lonsdale and make one final quotation that I found in her book 'Is peace possible?' (9). On p. 13, she is describing the hopes of the time after the Second World War, namely, for a 'peace settlement that would end all war. It might take time.... Meanwhile my work was fun. I often ran the last few yards to the laboratory. Later on I took my mathematical calculations with me to the nursing-homes where my babies were born; it was exciting to find out new facts'.

I ride my bicycle to the Lab, rather than running, but I also have the joy of being able to be enthusiastically leaping to my computer to undertake my crystallographic calculations that I can now do even on my laptop computer, whether at home or at the Lab. I feel the same as Kathleen Lonsdale about finding out new facts from the power of crystallography as an analytical tool, which appeals to the chemist and molecular biologist in me, as well as the grace and beauty of the foundations of crystallography, and diffraction physics, as a subject discipline, which appeal to the physicist in me.

I feel certain that Kathleen Lonsdale would have greatly appreciated a laptop computer. She would also have appreciated the wonderful capabilities of the current state-of-the-art and evolution of the electron microscope, neutron beams and synchrotron X-radiation as well as the prospects offered by X-ray lasers! Let us now move to part 2 of this article, SR in crystallography.

4.4 SR in crystallography

This Lonsdale Lecture 2011 describes the evolution and impact of SR X-ray beams in crystallography. A simple list of critical, advantageous, properties of SR X-rays I would give as

- Small focal spots
- Collimated X-ray beams onto the crystal
- Tuneable

And for special applications

- White beam for time-resolved fast (Laue) data measurements
- Defined time structure, for example picosecond bunches, for sub-nsec time-resolved studies
- Plane polarized

Clearly, a very promising list of X-ray beam properties for a myriad of applications in crystallography!

Synergies in methods developments and applications with neutron crystallography will also be described.

4.4.1 *First experiments and UK planning for SR protein crystallography; a personal perspective*

My first experiment with such X-rays was in 1976 at the Daresbury Laboratory, on the Synchrotron Radiation Facility (SRF) on the NINA high-energy physics synchrotron run by the UK's Science Research Council (SRC). The SRF had been set up under the leadership of Professor Ian Munro of Manchester University. My local scientific contact, employed by the SRC, was Dr Joan Bordas.

He made early pioneering developments on NINA (e.g. see ref. (*16*)). Joan is now Director of the Spanish SRS 'ALBA', which has been constructed under his leadership in Barcelona. (He retired in 2013 and was succeeded as Director of ALBA by Dr Caterina Biscari.)

I had first learnt of synchrotron X-rays in 1975 *via* a copy of a letter given to me by Prof. Dorothy Hodgkin: Dorothy was Nobel Laureate in Chemistry (awarded 1964). This letter she had received was from Professor Ron Mason to Professor Sam Edwards, Chairman of the SRC (*17*). Prof. Mason was visiting the Stanford Synchrotron Radiation Laboratory (SSRL) and where he had himself learnt of the first experiments with synchrotron X-rays in protein crystallography led by Prof. Keith Hodgson of Stanford University's Chemistry Department. This letter was accompanied with a preprint of an article finally published in *PNAS* in 1976 (*18*).

The text of this letter is reproduced below with permission of Prof. Mason. NB: SLAC is the Stanford Linear Accelerator Centre.

From Ron Mason, Visiting Professor of Chemistry, College of Chemistry, the University of California, Berkeley, California 94720
July 2 1975
To Professor Sir Samuel Edwards, F.R.S,
Chairman, Science Research Council, State House, High Holborn, London WC1, England

Dear Sam:
It is obvious that I cannot keep away from S.R.C matters! But I do have an interesting tale to tell as a result of spending a few hours on Sunday at SLAC and particularly at the synchrotron facility.

There is no need to emphasize the work that is going on in the general area of photoemission from metals, semiconductors and so on. Lots of valence band and core levels are being studied and there is also some nice work on reflection from metals using radiation energies down to 5 eV. Just coming on line is an instrument for measuring the angular dependence of photo-emission, something all of us interested in the field regard as very important indeed. There is also some X-ray absorption edge work going on, looking at transition metal complexes and metalloproteins but I was not impressed.

The main point I want to make, however, is this: you will remember that in our discussions of the S.R.F there was less than an enthusiastic response from crystallographers and I, quite frankly, played down their views for they are not known in general as a forward looking collection of people. Be that as it may, we heard particularly that it had no future in protein crystallography – that radiation damage effects would be impossible and that what was wanted was higher flux conventional fine focus X-ray tubes. It was not clear at the time how this view had been put together but I can now tell you it is nonsense! There has been three months work at SLAC on the proteins, rubredoxin and azurin. With a neat monochromator arrangement – after primary beam splitting, the beam hits a glass plate at grazing incidence (the plate is oriented by stepping motors) and then monochromatized by a single crystal of silicon or quartz. Complete data to 2.5 Å for a single zone is taking 2 hours; complete three-dimensional data are needing 60 hours of conventional precession photography; rubredoxin was still going strongly after 200 hours of exposure to different wavelengths and some very nice results on anomalous dispersion have been obtained. Azurin is going well after 150 hours exposure – typically an hour's setting photograph in the laboratory showed nothing, 2 minutes Polaroid photography using synchrotron radiation was enough to show the crystal orientation.

This is all, of course, based on a parasitic 20 mA beam; we should do much better. I think we should plan now and not accept the poor advice we received last Autumn. David Phillips, Uli Arndt and others can see this letter if you felt it appropriate as, of course, should Alick Ashmore (Director of Daresbury Laboratory) and Joan Paton. I am getting some working drawings of the monochromator and the crystallographers can get more details from Dr. Keith Hodgson.

I hope all goes well.

Yours,
Ron
Ronald Mason
Visiting Professor of Chemistry

Dorothy asked me my opinion. I reported to Dorothy that what was described sounded like the sort of direction we need to go in and that I would look into it in detail. It really buoyed me up to be able to talk to Dorothy Hodgkin herself like that! I was a DPhil student under the supervision of Dr Margaret Adams, a member of the Laboratory of Molecular Biophysics and Chemistry Tutor of Somerville College, and we were located in the Department of Psychology, Oxford University, in offices proximal to Dorothy's, as well as to Guy and Eleanor Dodson, and also John and Sue Cutfield. I was a very lucky DPhil student being in such an environment! I found reference 1 of Phillips et al. (1976) (*19*): namely the article Rosenbaum, Holmes and Witz 1971 *Nature* on 'Use of SR in biological diffraction' (nearly entirely on muscle fibre samples but with a mention of protein crystals as a possible sample in a summary table) (*19*). I reported to Dorothy that I found this a most promising development as, in my view, it got round major limitations in laboratory methods in X-ray protein crystallography of the time. It was suggested that I meet with Professor David Phillips, who was the Head of the Laboratory of Molecular Biophysics, based in the Zoology Department of Oxford University. He had interviewed me in the spring of 1974 for a DPhil place in his Laboratory, which was the overall organizational grouping for protein crystallography in Oxford. I had requested to be supervised by Dr Adams on her particular project (X-ray crystallography studies concerning the structure of the enzyme 6-phosphogluconate dehydrogenase (6-PGDH)), which was nicely at the beginning of the crystallographic stage of the project, I felt. Prof. Phillips, in response to my request to visit the SSRL to learn more, after my meeting with Dorothy, suggested instead that I submit a proposal to do the experiment I had in my mind to the SRC NINA SR Laboratory at Daresbury Laboratory. He asked me if Warrington was a problem, compared with Stanford. I said 'No, Warrington would be fine, I know they have a good rugby league team', he laughed. The experiment I wished to undertake was to optimize the anomalous scattering at the Pt LIII absorption edge *via* maximizing the f'' value for the 6-PGDH's $K_2Pt(CN)_4$ heavy atom derivative, which I had found in my first year's DPhil research. I undertook a 24-h beamtime run at NINA with Dr Joan Bordas in April 1976 shortly before NINA closed. I described this experiment in my DPhil thesis. It was not successful in terms of results, but it was important to me for learning about the beamline equipment that I would need to carry the experiment further. This did become possible when I took up an appointment in February 1979 at Keele University 50% jointly with Daresbury Laboratory to develop my ideas (*20*). This led to the first SRS instruments for protein crystallography, serving users as well as developing new methods, namely, SRS 7.2, 9.6 and 9.5. I became full-time at Daresbury at the cessation of my appointment in 1983. The user support component grew to be too much and I decided to join York University on a Joint Appointment in late 1985. This continued (including a move to Manchester University in January 1989) until mid-1993, when I became increasingly pre-occupied helping to develop ESRF's macromolecular crystallography plans through the mid-1980s and 1990s.

In 2011, I got further historical insight into the UK position regarding SR protein crystallography when I came across the following document:

'The Scientific Case for Research with Synchrotron Radiation' SRC Daresbury Lab DL/SRF/R3 (1975) (*21*). Within this was the following text:

'Professor Phillips' Molecular Enzymology Group at Oxford is also very interested in the possibility of using synchrotron radiation to study transient structural changes in crystalline enzymes. The availability of more intense X-ray beams will also make it possible to extend present day crystallographic structure determinations to molecules with much larger unit cells and to ones which can only be obtained as very small crystals – provided that means can be found to reduce radiation damage. Present studies indicate grounds for cautious optimism in this respect. It seems likely that when the feasibility of carrying out diffraction experiments within very much reduced periods of time become more generally recognised, there will be no shortage of applications in other fields.

For reasons which are discussed elsewhere in this report most X-ray diffraction work on biological material is done, and will continue to be done, using wavelengths in the wavelength 1.5 Å region. Thus a synchrotron or storage ring designed to give an output in the wavelength 1.5 Å region would be a very suitable source. Wavelengths longer than this could be used only with a substantial loss of intensity in most specimens. Shorter wavelengths are unlikely to be required'. Obviously, my ideas in ref. (20) were partly already covered by David Phillips's input in 1975: I say 'partly', because wavelengths shorter and longer than 1.5 Å it seemed to me were already very clearly going to be important for optimizing anomalous scattering applications in protein crystallography. Also short wavelengths would allow improvements in diffraction data accuracy and longer wavelengths would increase the sample scattering efficiency such as for small crystals.

4.4.2 *The Daresbury SRS and the start of the so-called second-generation SR sources*

The SRS at Daresbury was the world's first dedicated SR X-ray source. The details of its genesis were before I became involved in the organizational politics of any such matters. I learnt at second hand that Dr Ian Munro of Manchester University had instigated the NINA SRF Facility and its success meant that the SRS proposal was well received and approved for funding. There is a document of plans for the SRS from 'The Crystal Optics Panel' (22). Listed in that Panel, as well as David Phillips himself, is Dr Michael Hart, a key pioneer of SR in crystallography, notably for example for his developments of perfect crystal optics and instrumentation development such as small-angle X-ray scattering and interferometry (see, e.g. (23–25)). Professor Hart later became Director of the USA Brookhaven National Laboratory's National Synchrotron Light Source (NSLS). Professor Munro became SRS Director, and was the person to turn off the SRS for the last time at a ceremony held at Daresbury in August 2008: Ian's review of the SRS up to the late 1990s is in ref. (26). I was Director of the SRS in 2002, until I returned to Manchester University in January 2003 to focus more on research than SR source administration.

In ref. (22) the case of 'Lattice structures' does feature. Thus, for instrumentation, the report states:

'*3.5 Lattice Structures and Interferometry*

Conventional X-ray sources are adequate for all but the most demanding lattice structure problems. It follows, therefore, that the SRS facility will be used by only those crystallographers who have problems where particular advantage can be gained from a free choice of wavelength with high intensity, low angular divergence, or high degree of polarisation of the primary beam. The monochromated beam will also reduce specimen damage. Apart from a few interested protein crystallographers, potential users are hard to identify. We can assume, however, that future application will include work with small or unstable crystals, and with specimens from which either accurate lattice parameter or structure amplitude measurements are required. Part of the programme will be concerned with the interferometric measurement of X-ray optical constants which are required for structure analysis.

Users of the SRS facility are likely to fall into one of two main categories:

(a) *those who require rapid collection of vast amounts of structure amplitude data;*
(b) *those who make relatively few measurements but require precise control of the experimental conditions.*

The experimental needs of the first group are relatively straightforward and typically satisfy the requirements of the protein crystallographer. Experiments performed by the second group are likely to be less routine in nature, and instrumentation specified for these users must be sufficiently flexible to accommodate unidentified demands. Taking into consideration these requirements and the expected use of the SRS facility, it is anticipated that the needs of lattice structure users will be satisfied by two beam line locations and the following equipment:

(i) *A low-resolution scanning monochromator covering the range 0.5 to 4 Å and producing a uniform beam of diameter 0.05 to 0.20 mm. The sample should be mounted on a crystal oscillation system similar in concept to that marketed by Enraf-Nonius, but with a large area detector and associated computer readout.*

(ii) *A high-resolution scanning monochromator covering the widest possible wavelength range, and producing a uniform beam of approximately 1 mm in diameter. The crystal sample should be mounted on a computer-controlled orienting system equipped with beam shaping facilities, a "small" area detector device and associated computer readout.*

The Panel recommends that two instruments be provided for work on lattice structures at the SRS, one of each type discussed above'.

For crystallographic computing, the report states:

'Users will also want to analyse their spectra online wherever possible but in some instances e.g. crystallographic work, the large-scale computing facilities that are required would probably have to be provided off-line'.

The other research and technique areas covered by the report were: *EXAFS (Extended X-ray Absorption Edge Fine Structure); XPS (X-ray Photoelectron Spectroscopy); X-ray Topography; SAS (Small Angle Scattering); X-ray Absorption Microscopy and Lithography;* finally a development programme including radiometry was also identified.

Clearly, this was a far-seeing report for the SRS planning, in which one can pick some holes today, but these crystal ball gazing exercises are not easy, as I was to wrestle with myself with the ESRF Macromolecular Crystallography Planning Group.

Where did SRS sit in the evolution of SR X-ray sources? Professor Michael Hart drew my attention to Figure 4.5 (prepared by Dr Gwyn Williams of Jefferson Lab, USA, and who kindly has let me use this here). As Michael would describe this, what better way than to show the profound change in X-ray source capability by comparing the SR source X-ray brightness along with computer cpu power, versus calendar year. The SRS emittance placed it in the 'second-generation' part of the graph, impressively more than previously but with considerable further optimizations yet to come with the third-generation, such as the ESRF and the fourth generation, the X-ray laser. A useful summary of the evolution of sources, and particularly projections in the early 2000s decade, is given in ref. (*27*).

4.4.3 *The SR research community gets SR News and then its own Journal of Synchrotron Radiation (JSR), the latter published by IUCr*

Around the early 1980s, I recall a visit by a commercial publisher representative canvassing opinion about launching a community newsletter, *SR News*. I recall supporting this enthusiastically. It has been published for many years and is very professionally produced and with full representation of the SR facilities and with senior SR scientists as Editors. As well as meeting reports, they carry excellent articles. I have submitted quite a few meeting reports, including two summarizing SR presentations at IUCr Congresses. As Chair of the IUCr Commission on SR, I gathered together performance data for the available SR protein crystallography instruments of the time and *SR News* accepted it for publication (*28*). I also, most recently, had my report of the 2009 BAAS Meeting

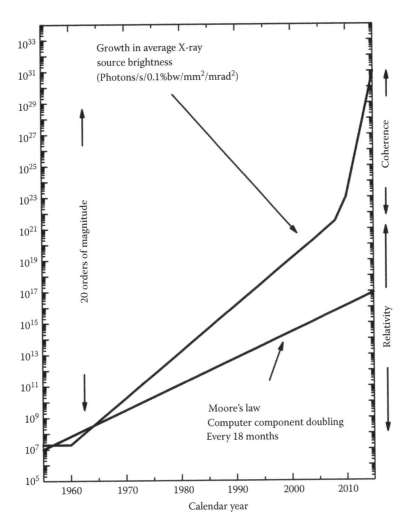

Figure 4.5 The evolution of SR source brightness vs. calendar year. Figure courtesy of Dr Gwyn Williams, Deputy Associate Director, FEL/Light Sources, Jefferson Lab, Newport News, VA, with permission of the author.

published in *SR News* in 2010, which has been well received, with a full lifetime look-back of the SRS and its science programme as it evolved over the years, as well as a look forward (*29*).

In the early 1990s, Samar Hasnain put forward the idea to the IUCr to launch a new *Journal of Synchrotron Radiation*. I was at that time the inaugural Chair of the IUCr's Commission on SR and was glad to support the idea. I vividly remember what seemed to me a pivotal meeting held at IUCr Chester in March 1992 with the IUCr Treasurer, Prof. Asbjørn Hordvik from Tromsø University, Jim King (IUCr Executive Secretary), Mike Dacombe (IUCr Technical Editor), Samar and myself. I tabled about five of my own SR instrumentation papers, which were in a variety of non-IUCr journals, including *Nuclear Instruments and Methods, Journal of Physics E: Scientific Instruments, Review of Scientific Instruments,* etc. I was asked by Asbjørn, 'Why did I not submit such papers to the *Journal of Applied Crystallography*?' to which I replied (i) I was constrained by where the SRI Conferences chose to publish their conference proceedings and (ii) I needed to reach a wide instrumentation audience. Some authors still prefer *J Appl Cryst* over *JSR* for their SR instrumentation and methods papers. Nevertheless, considering the wide spectrum of SR activity, it was soon recognized widely in the IUCr community that *J Appl Cryst* was not the best journal for publications in fields only loosely connected with crystallography, such as those arising from the vacuum ultra-violet (VUV) community. Indeed, the scope of *JSR* was set to include such techniques as well as of spectroscopy and microscopy, and in 1993, we gave a report on our business plan at the IUCr Congress in Beijing (*30*).

The business plan was discussed at a formal meeting of the IUCr Finance Committee on 21–22 March 1992 in Chester, United Kingdom, with the full attendance being R. Diamond (Convener), A. Authier (President), A.I. Hordvik (General Secretary and Treasurer), C.E. Bugg (Editor-in-Chief), S.G. Fleet (Investment Adviser), A. M. Glazer (for the discussion of *JSR*), M.H. Dacombe (IUCr Journals Technical Editor), J.N. King (IUCr Executive Secretary) and A. Cawley (Secretary) plus Samar and myself for the item on the discussion of *JSR*. The business plan was approved by the IUCr Executive Committee under the Presidency of Prof. Andre; Authiér in 1993, obviously following support from the IUCr Finance Committee. The Journal was to have three Main Editors: Samar, myself and Prof. Hiromichi Kamitsubo, then Director of Super Photon Ring 8 GeV (SPRing-8), Japan.

The Journal was launched with its first issue in October 1994. Our Editorial (*31*) stated that:

The IUCr has taken this initiative because the diffraction community has a strong vested interest in synchrotron radiation, and therefore in harnessing the best features of synchrotron radiation instrumentation and methods. We, and the IUCr, believe that the full benefit of this initiative can only be felt if the *Journal of Synchrotron Radiation* serves the whole of the synchrotron radiation community, across its full spectrum, rather than covering the hard X-ray region alone. This diversity is reflected in the format and scope of the journal and was ensured by conducting the widest possible consultations. Thus, we have approached synchrotron radiation representative organizations and directors of synchrotron radiation facilities, and have made presentations and solicited comments at a wide variety of synchrotron radiation conferences. The Editorial Board reflects this breadth of representation and provides wide-ranging coverage of the interests of the synchrotron radiation community.

The inaugural front cover shows the proliferation of SR sources and thereby the major expansion of this research tool that had already taken place (Figure 4.6). Estimates of the crystallography usage of the SR X-ray sources were made to be around 50%. The viewpoint of the IUCr Executive Committee with respect to this initiative, and the detailed timeline of events leading to approval for launch, including the balancing of the significant risks and opportunities, as well as some opposition to the new journal, is described in the article by André Authier (*32*). In my Triennial (1990–1993) Report to the IUCr Executive Committee as Chairman of the Commission on Synchrotron Radiation (*33*), I reported on this Journal initiative as follows: '*A proposal for a Journal of Synchrotron Radiation has been made to the IUCr. The Commission has had a major input to the deliberations on this. Open discussions were held at a variety of conferences, both national and international, in Europe, the USA and Japan. Representative organizations and the Directors of Synchrotron Radiation Laboratories have also been contacted*'. It was indeed a great deal of hard work getting *JSR* approved and launched, as well as the first 5 years of operation when I was one of the Main Editors. In these early years, there were regular headaches about thin issues of the Journal (*34*) and regularly firm questioning by the IUCr Finance Committee, rightly so, which I had to field as IUCr Journals Editor-in-Chief 1996 to 2005.

The evolution of the Journal, and the firm financial commitment of the IUCr, can be judged from our Editorials between 1994 and 2000 (*31, 34–36*). As we wrote in our Editorial in 1996 (*34*):

As we enter our third year and write this editorial, we have received the first major review of the *Journal of Synchrotron Radiation* (JSR) (*37*). It is worth repeating some of the points made in this

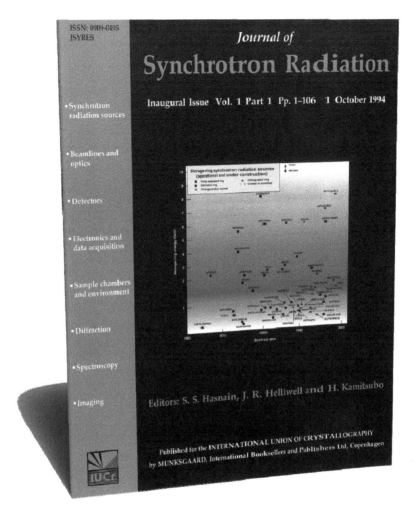

Figure 4.6 The first front cover of the *Journal of Synchrotron Radiation*. Published by IUCr, with the permission of the IUCr.

independent review. It says that 'developments in the application of synchrotron radiation research have benefited enormously from fertilization between otherwise distinct research areas, and there is no doubt that this new journal will play an important part in furthering such interdisciplinary interactions'. The review goes on to say that the 'speed, together with the quality of the contributions so far and the high standard of production, makes the journal attractive to authors and required reading for workers in what worldwide is still a rapidly expanding field'. We note that the reviewer points out that even though the quality of articles in *JSR* has been high, the issues have remained fairly slim.... Our main focus now is to increase the size of the issues without compromising the quality of the papers.

IUCr had given firm financial commitments as we wrote in 1995 (*35*):

As we enter into our third volume we feel confident that *JSR* is here to stay, as it is essential for the continued growth and stimulus of synchrotron radiation facilities, techniques and applications. But we approach a crucial period. This is the last complimentary issue of *JSR*. Thus, it is essential that we call upon you to ask your libraries and institutions to place subscriptions immediately.

In 1999, we were able to write (*36*):

> With this issue, we celebrate the fifth anniversary of the journal. Since the launch, approximately 850 papers and 3800 pages have appeared. The journal has published the proceedings of two main synchrotron radiation conferences, SRI'97 (May 1998 issue) and XAFS X (May 1999 issue), where new standards for these proceedings have been set. The journal now features in the top 17% of the Science Citation Index (4800 journals). Its impact factor is greater than that of *Rev. Sci. Instrum., Nucl. Instrum. Methods, J. Phys. A* and *J. Phys. C,* and is approaching that of *Phys. Rev. C* and *Phys. Rev. E.* Thus, the *Journal of Synchrotron Radiation* (*JSR*) has become clearly established and this it owes to the confidence the community has placed in it from its launch.

In 2011, one can observe and congratulate the current Main Editors, Gene Ice, Åke Kvick and Toshiaki Ohta, that *JSR* is enjoying regular thick issues and also a pleasing buoyancy of participation of quite a few of the SR facilities with sponsored 'Facility pages'. The latest obvious major step in the *JSR*'s evolution has been the addition of the logo on the Journal front covers of 'including free electron lasers', recognized by the *JSR* Editorial in 2001 (*38*).

4.5 SR theory and actual magnet sources

The details of the theoretical physics of this phenomenon were originally derived by Prof. Julian Schwinger (*39*), and a history of the first demonstrations of SR is described in ref. (*40*). For the purposes of illustration of such matters, I offer the following examples.

The SRS bending magnet shown in Figure 4.7 is a 1.2 T magnet, which for the SRS, a 2 GeV machine has an SR universal spectral emission critical wavelength of 4 Å. Thus, SRS Station 7.2, fed by SR from such a magnet source, provided useful flux for experiments and data collection in protein crystallography from a minimum of ~1.2 Å. The maximum useful wavelength on SRS 7.2

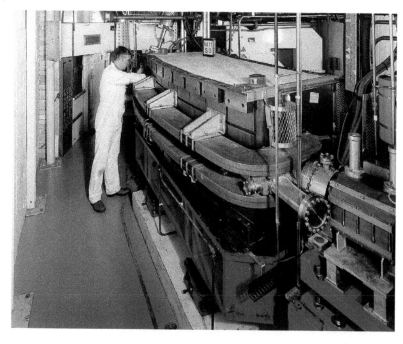

Figure 4.7 An SRS bending magnet (1.2 T magnetic field) reproduced with the permission of UK Science and Technological Facilities Council (STFC) Daresbury Laboratory.

was approximately 2.6 Å. The longer wavelength limit was set by beryllium windows in the beam-line, which caused X-ray absorption and a steady reduction of X-ray intensity from ~2.0 Å onwards up to a limit of 3 Å. SRS 7.2's usual range of wavelengths for data collection spanned monochromatic values from 1.38 Å up to 1.86 Å, and most typically of 1.488 A for users.

The SRS had straight sections which could accommodate insertion device magnets of various types. The ESRF was designed to be mainly an SR source harnessing insertion devices as light sources, rather than the bending magnets. The angular emissions for a single bump wiggler magnet, a multipole wiggler (MPW) and an undulator are shown in Figure 4.8 along with magnets for each type, as illustrated from SRS and ESRF. The SRS superconducting wiggler (SCW) type of magnet in Figure 4.8, of which two were installed in the SRS (in Sections 9 and 16), allowed a much higher magnetic field than the 1.2 T bending magnets. Thus, line 9 at 5 T had a critical wavelength of its emission of 0.9 Å. The SCW 16 with 6 T had an emission critical wavelength of 0.75 Å. The electron beam path through an SCW is shown in the schematic top diagram. The natural emission cones of light have opening angles much smaller than the angular deflection through the straight section of the SCW magnet's field.

The middle diagram and middle photo below it are for a MPW. These magnets have a value of around 2.5 T, with a critical wavelength of emission for the SRS 2 GeV of ~2 Å. The electron beam excursion through this straight section's magnetic field is not as marked as through the SCW, but the SR cone angle at each peak and trough of the electron beam's travel do not overlap. The X-ray emission is a simple multiple of the number of poles of a bending magnet of equivalent field: typically 10–20 times increase of emitted X-ray flux is achieved over a 2.5 T (hypothetical) bending magnet. The bottom diagram of Figure 4.8 shows the undulator case. The weaker magnetic field is chosen so that the emission of the electrons, as they pass through the magnet, are deliberately allowed to overlap. Thus, there is a constructive interference condition and the polychromatic emission spectrum collapses into a few spectral lines but of duly increased brightness. The brightness increases observed are up to N^2, where N is the number of poles in the magnet; so with N typically up to 100, gain factors of X-ray brightness are 10^4, i.e. colossal. The ESRF undulator type is shown: the ESRF at 6 GeV has a high enough machine energy to have undulator emission into the X-ray region. Indeed the ESRF was initially conceived as a 5 GeV machine, which with a 20 mm magnet gap would have an undulator fundamental emission of 1 Å wavelength. In order to provide a fundamental at 0.8 Å for nuclear resonance experiments that began to be proposed, the ESRF machine energy was increased to 6 GeV. An additional requirement for the ESRF undulator interference condition is to have a small electron beam size. Thus, with the high machine energy of 6 GeV, which reduces the SR natural opening angle, the machine emittance is very small (i.e. yielding a high machine brightness).

The ESRF then was conceived with a much improved emittance, i.e. smaller than the SRS, as well as a much larger circumference to allow many and long straight sections. Figure 4.9a shows the SR universal curves for the ESRF for each magnet type. (It has two bending magnet types with the magnet fields shown.) The curves for the undulators in Figure 4.9a are the tuning ranges possible by altering the magnet pole pieces gap (20 mm to larger). Figure 4.9b shows an undulator emission for a given ESRF magnet pole-piece gap. This is the ESRF 6 GeV case for an undulator's X-ray emission for the details given in the inset, i.e. two different cases (at the time of the Foundation Phase Report in 1987 (*41*); these have improved considerably since then, notably in terms of spectral brightness). Narrower magnet gaps are possible today (a pioneering example was at the NSLS, see ref. (*42*), which reached 3.8 mm gap with 3 mm gap indicated as possible). This is important as it explains why the lower machine energies of Diamond Light Source (DLS), Soleil, etc. (3 GeV) still yield useful X-ray emission from a harmonic at 0.98 Å wavelength, the selenium K edge, a popular wavelength amongst macromolecular crystallography (MX) users today.

My earliest experience of the undulator magnet concept in practice was from the visit of a Soviet science delegation to Daresbury in approximately 1984. Not long after this visit an

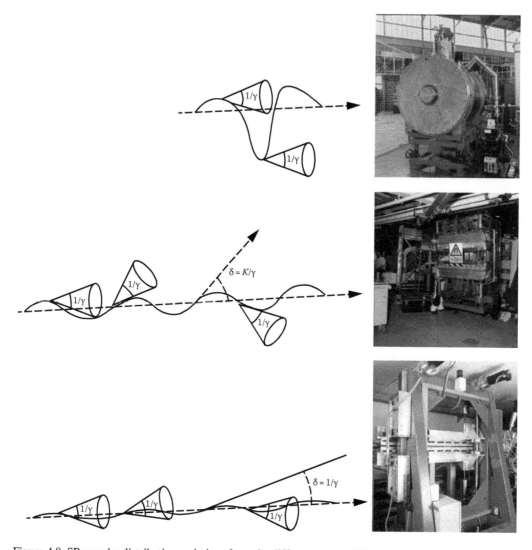

Figure 4.8 SR angular distribution emissions from the different types of SR sources shown: Top: the super-conducting wiggler (SRS SCW), Middle: the multipole wiggler (SRS MPW), Bottom: the undulator (ESRF type). Line drawings reproduced with the permission of Cambridge University Press from ref. (*103*) and the photos reproduced with the permission of the STFC Daresbury Laboratory (top, middle) and the ESRF (bottom). K is the undulator parameter, δ is the angular deflection of the electrons in the magnetic field and γ is the electron's energy divided by its energy at rest.

undulator was installed on SRS machine lattice straight section 5. I submitted the colour picture of its visible light emission for the front cover of a special issue of *Chemistry in Britain* (now called *Chemistry World,* published by the Royal Society of Chemistry) on SR in the chemical sciences (my article with Prof. C.D. Garner is ref. (*43*)). This emission, being in the visible region of the electromagnetic spectrum, showed nicely the SR emission interference condition, being both wavelength and angular position dependent; thus, the use of a pinhole suitably placed can be a simple means of selecting appropriate portions of the emitted fundamental and harmonic interference peak wavelengths.

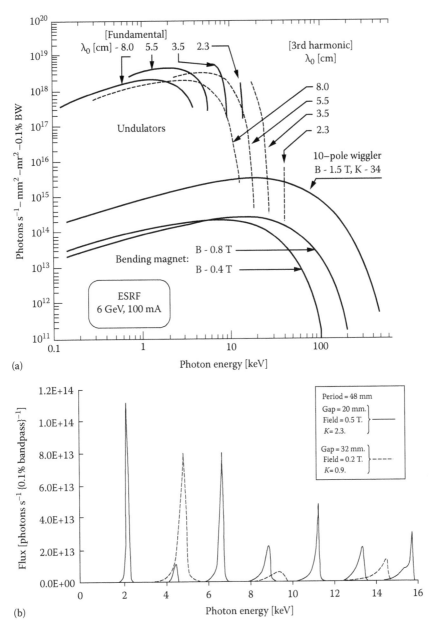

Figure 4.9 (a) The SR spectral curves for the ESRF. (b) The ESRF undulator emission for the two conditions shown in the inset. Full line: Magnet gap = 20 mm, field = 0.5 T, K = 2.3. Dotted line: Magnet gap = 32 mm, field = 0.2 T, K = 0.9. Both plots are for the ESRF machine performance as at the time of the ESRF Foundation Phase Report in 1987 and these have improved considerably since then, notably in terms of spectral brightness. Reproduced with the permission of ESRF.

4.6 The importance of the second- and third-generation SR sources in bringing about a revolution in crystallography

4.6.1 *SRS Station 7.2: the first protein crystallography instrument on a dedicated SR X-ray source*

The SRS was a dedicated X-ray storage ring and, unlike the NINA synchrotron, allowed for dedicated control for SR users from the SRS Machine Control Room of the injection and beam position and stability, albeit not with source position feedback control immediately. The first station/instrument for protein crystallography at the SRS was on bending magnet 7, 'SRS 7.2'. This beamline, Number 7, was the first X-ray beamline for SRS and was developed in parallel with beamline 6 for VUV research. X-ray beamlines were separated from the main machine vacuum vessel by a beryllium window, whereas the VUV beamline design had a progressive change in quality of the vacuum, with the machine being held at the best vacuum. A sequence of fast valves was in place on the VUV beamline in the case of a sudden breach of gas into the beamline and thereby the machine. The competition between research communities to secure a place on these first beamlines was significant and so, an essential first step I took was to get the backing of the UK protein crystallography ('PX') community behind a beamline bid. Thus, I called a meeting held at Daresbury of the UK PX research group leaders. Support for the concept of SRS PX 7.2 was secured and I presented the X-ray optical design at the SRS User Meeting of September 1979. I was helped significantly here by the experience of Mr Jeff Worgan of Daresbury Laboratory and indeed, we prepared a Technical Memorandum on the monochromator design *(44)*. Initially, the idea was to have a point-focussed beam from a doubly curved monochromator, but we concluded that the design *(45)* at Laboratoire pour l'utilization du rayonnement électromagnétique (LURE), Paris, and Deutsches Elektronen-Synchrotron (DESY) Hamburg of a triangular-shaped, Fankuchen cut perfect Ge *(111)* monochromator would produce a 10:1 demagnified horizontal source, which was 14 mm for SRS, down to 1.4 mm and thus approaching the expected typical protein crystal sample size of 0.3 mm. A further challenge, besides the very large horizontal source size, in the design for SRS 7.2 was to secure a focussed vertical beam and which could take advantage of a rather nice small SRS vertical source size of 0.3 mm. Dr Richard Tregear, a fibre diffractionist, offered to help by asking Prof. Ken Holmes, ex-Director of the European Molecular Biology Laboratory (EMBL) Outstation Hamburg, for assistance. Prof. Holmes responded with actually letting us have one of his original double 20 cm reflecting quartz mirrors and bender! On detailed consideration however I decided I liked the competing approach pioneered at SSRL of a single-segment platinum-coated fused silica mirror of 60 cm, which is easier for the mirror bender and alignment. This would accept the full vertical divergence of the SRS 7.2, as well as the 4 milliradians (mrad) of horizontal divergence beam allocated to us. China Lake, in the United States, supplied the platinum-coated mirror. The perfect germanium crystals for the monochromator came from a company in Grenoble at the French Atomic Energy Authority 'CEA' called 'Cristal Tec' run by Monsieur Guinet, whom Jeff Worgan knew. I ordered these monochromator crystals with a 10° Fankuchen cut and not long afterwards, I also ordered the Si *(111)* Fankuchen cut crystals, but this time with several cut angles to allow for several X-ray wavelength ranges to be harnessed. As part of the acceptance tests for their quality, I took them all to Durham University to work with Prof. Brian Tanner's research group (including Dr Graham Clark, who later joined the SRS, and Ms Siti Gani, a PhD research student there) to undertake double-crystal diffractometer rocking curve scans, which took a week, and which were fine.

The whole SRS 7.2 mechanical arrangement for the instrument was a big design and build effort that was led by Mr Phil Moore, the mechanical engineer assigned to work with me by the Daresbury Laboratory. There were the various mechanical assemblies and motors controlling all alignment of apparatus, of the mirror bender, of the monochromator rotation stage and the concomitant experimental arm of the instrument, which carried the protein crystallography apparatus (initially an Arndt Wonacott rotation film camera). The mirror bender and the camera alignment table were built to Phil's design in the Keele University Physics Department Mechanical Workshop. Between

1979 and 1983, I was a Joint Appointee between Keele University Physics and the Science and Engineering Research Council (SERC) Daresbury Laboratory (50%/50%). I also obtained a research grant from SERC to work on the *Bacillus Stearothermophilus* 6-phosphogluconate dehydrogenase, whose crystals were a gift from my DPhil supervisor Dr Margaret Adams. This grant included the development of the SRS 7.2 station and funded a post-doc, and thereby, I was joined by Dr Trevor Greenhough, a very experienced crystallographer. My first two PhD research students, Mr Paul Carr and Mr Stephen Rule, took courses which I gave in the Physics Department. Trevor and I were very interested in the data processing of the oscillation camera films we would measure at SRS 7.2. Trevor, with Paul Carr and help from Daresbury, wrote the software for the SRS 7.2 motors' control *via* a controller box designed and built by Daresbury. Trevor also started deriving new formulae for the prediction of the protein crystal reflecting range and partialities important for high-quality data estimations and, in the case of virus crystallography, essential as only one diffraction film could be measured per crystal; thus, there were many partially recorded reflection intensities which had to be suitably scaled. As I was commissioning the SRS 7.2 instrument, and thereby understanding the basic properties of the SR X-ray beam spectral spread and convergence angles from the X-ray optics scheme I had adopted, it became clear to Trevor and me that a new formalism of the oscillation film camera data-processing software for partiality correction was needed. What's more, these details had not appeared in the literature before. Thus, we obtained our first and indeed well-cited articles in *J. Appl. Cryst.* on this topic (*46, 47*). The SRS 7.2 instrument paper was submitted to a scientific instruments journal and at that time, the UK Institute of Physics Journal *J. Phys. E: Sci. Instrum.* was suitable. This paper could also be cited by users of the instrument, since I did not feel that my (automatic) co-authorship of articles published by users was appropriate. This paper (*48*) also explored and calculated the focus at the crystal or at the detector options and, for these various reasons, became very well cited. We also had a paper accepted in *Nature* (*49*) in a collaboration with Dr Uli Arndt from the MRC Laboratory of Molecular Biology, Cambridge, which involved the over-bent setting of the SRS 7.2 monochromator to create a polychromatic profile in each and every diffraction spot. This was demonstrated using a test crystal of a rhenium compound provided by Prof. Judith Howard; this technique later became known as diffraction anomalous fine structure (DAFS).

The properties of the single-bounce monochromator as an X-ray optic fascinated me. So I set about making practical tests of the monochromator in different states of curvature and also examining the spectral behaviour of the horizontal focus. I had pressed Phil Moore to give me a wide rotation range not only for the monochromator but also for the beryllium window in the monochromator vacuum vessel, even allowing me to examine the reflected X-ray beam from the monochromator at $2\theta = 90$ degrees. The instrument 2θ arm went up to approximately $60°$.

Figure 4.10 shows a schematic diagram of the range of incident angles across the monochromator crystal surface (from ref. (*50*)) and which thereby reflected particular wavelengths in the directions shown. The required setting for usual data collection is as shown in Figure 4.10b which minimized the spectral spread of the X-ray beam incident onto the crystal sample. For different wavelengths, the focussing distance for this minimum spectral spread setting had to be varied because of the Fankuchen cut angle and so, the oscillation camera and its alignment table had to slide along the length of the 2θ arm. In Figure 4.10c is shown the overbent monochromator condition and where a simple correlation of incident angle with wavelength could be set up onto the crystal and so create the polychromatic profiles mentioned above (*49*). In Figure 4.10d is shown the curvature condition Figure 4.10b, but now illustrating the wavelength correlation across the focal width; thus, by placing a narrow slit in the focus and simply translating it across the focal full width, one could select specific wavelengths and thereby make an edge scan such as from a nickel foil.

For the monochromator 2θ scan through $90°$, mentioned above, I had to set up a simple table to hold an X-ray film cassette and so produced a trace of the monochromator reflected X-ray intensity; at $90°$ of course, the trace showed a minimum intensity due to the nearly perfect linear polarization of the synchrotron X-ray beam. The minimum intensity was not zero as there was a residual, small, intensity from the vertical component of the beam polarization. All these details

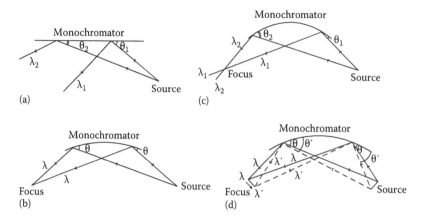

Figure 4.10 The single perfect crystal as monochromator for use with SR in its various settings. (a) For a point source and a flat crystal there is a gradual change in the photon energy selected from the white incident beam as the length of the crystal is traversed. (b) For a point source and a curved crystal, one specific curvature can compensate for the variation in incidence angle. The reflected spectral bandwidth delta λ/λ is then, of course, a minimum and this setting is known as the Guinier position. (c) If the curvature of the crystal is increased further a range of photon energies is selected along its length so that the rays converging towards the focus have a correlation of photon energy and direction. (d) A finite source causes a change in incidence angle at the crystal so that at the focus there is a photon energy gradient across its width (for all curvatures). Placing a slit at the focus to reduce this delta λ/λ contribution onto the protein crystal is akin to placing a slit at the tangent point of the SR source. Use of such a slit is at the expense of flux; with a high brightness source (small source size) such a slit is not needed. From ref. (*50*) with permission.

formed part of the Helliwell et al. *J. Phys. E* paper (*48*) documenting that the instrument's X-ray optical properties were well understood.

These details fuelled the various ideas that Trevor and I had about their impact on the oscillation camera data processing, with Trevor undertaking many formulae derivations that feature in our article (*47*), following on from the preceding article that we published for conventional X-ray sources (*46*). We also published a more detailed technical paper (*51*) to accompany the *Nature* article (*49*). With our aims being mainly at that time protein crystallography, our efforts to convert the development of ref. (*49*) into protein structure applications proved too difficult due to the limited detector quantum efficiency (DQE) of film to convert a spot intensity into a measured spectrum with weak intensity variations that would accrue from a dilute metal atom in a protein. This was a pity because to move from a single-wavelength diffraction data set, albeit with the marvellous property of being fully tuneable, to a full spectrum in each and every diffraction spot width, was of course in principle very attractive. Much later, other groups independently rediscovered the approach and named it DAFS and diffraction anomalous near edge structure (DANES) and applied it in inorganic chemistry and materials science crystallography (see, e.g. (*52*, *53*)).

The first users of SRS 7.2 of course included all the UK research laboratories undertaking protein crystallography at the time. Prominent in my memory were: Oxford Molecular Biophysics (David Phillips, Louise Johnson, Janos Hajdu); my DPhil supervisor Margaret Adams who provided 6-phosphogluconate dehydrogenase crystals from which to collect data; Birkbeck College (Tom Blundell, Peter Lindley); the York Structural Biology Laboratory (Guy Dodson, Zygmunt Derewenda, Bob Liddington); the Sheffield Laboratory (Pauline Harrison, David Rice, John Smith); the Imperial College Blackett Laboratory (David Blow, Alan Wonacott); Bristol Biochemistry (Herman Watson, Hilary Muirhead); MRC Laboratory of Molecular Biology, Cambridge (Anne Bloomer, Phil Evans); as well as the nucleic acid crystallographers from Olga Kennard's Laboratory at the Cambridge Crystallographic Data Centre. An early and prominent methods collaborator was Uli Arndt from the MRC Laboratory of Molecular

Biology, Cambridge, as mentioned above. We also were becoming noticed internationally: Michael Rossmann arrived promptly from Purdue University with his human rhinovirus crystals; Howard Einspahr (pea lectin) and Steve Ealick (purine nucleoside phosphorylase) from Birmingham Alabama. From Sweden came Carl Branden, Anders Liljas and Ylva Lundquist from Uppsala, and they subsequently established a formal agreement with SRS regarding beam time access that lasted for many years, including leading to SRS 9.5 (see below). Several of these users were first-time users, although others had already visited the LURE and/or EMBL Hamburg instruments for protein crystallography.

Michael Rossmann's work at SRS 7.2, as well as his work at EMBL Hamburg, led to a protocol for virus crystal data collection which was called 'the American Method': 'shoot first and ask questions later'. This R&D Michael wrote up, with his co-worker John Erickson, in *J. Appl. Cryst. (54)*.

An early policy issue was regarding publications by our users and our expectation regarding our role in the work done, and this was a question raised by our visitors as well. I have no memory of guidance from the SERC management, but I was contacted by Dr Roger Fourme from the equivalent protein crystallography facilities at the French LURE synchrotron and by Dr Hans Bartunik from the EMBL in Hamburg. I established a policy at SRS 7.2 where we would publish an instrument paper (i.e. *(48)*), which general users of SRS 7.2 would be requested to cite, as the place where they had collected their data. Collaborations however could naturally lead to co-publication, and that is what happened.

As well as the basic characterization of the SRS 7.2 monochromator, I also set about exploring the wavelength range capabilities of the instrument. Phil Moore, the mechanical engineer, had worked hard in providing a very versatile instrument 2θ arm, which allowed me to explore short and long wavelengths. I used metal foils and a pair of incident beam and transmitted X-ray beam ionization chambers to measure the X-ray absorption spectra and thereby to locate elemental K edges. Thus, the zinc (1.28 Å), copper (1.38 Å), nickel (1.488 Å), cobalt (1.608 Å), iron (1.743 Å) and manganese (1.896 Å) K edges were explored. Diffraction data were most conveniently collected at 1.488 Å wavelength, for which the Guinier condition, monochromator minimum spectral bandpass, was easily established in the mid-range of the allowed focussing distance on the 2θ arm. Away from this wavelength, it would become necessary to reduce the horizontal acceptance of the monochromator from the available 4 mrad so as to control the spectral spread values; this was done with the pre-monochromator horizontal slits, which were under motor control. I undertook collaborative experiments with Dr Howard Einspahr on the pea lectin protein at the manganese K edge *(55)* to locate the manganese and calcium ions in the protein dimer using the optimized Mn anomalous differences. The crystal sample was aligned very carefully so that each rotation photograph had anomalous differences left to right across each film, with a view to minimizing any impact of radiation damage or SRS incident beam decay and fluctuations on the small anomalous intensity differences expected (see Figure 4.1 of ref. *(55)*).

A quite speculative set of trials was going to much longer wavelength and I reached 2.6 Å. I think I used a vanadium foil to calibrate the wavelength to 2.269 Å and then made a simple extra calculated θ change to set the desired 2.6 Å wavelength. This value I had decided upon because of the simple practical limitations of my various beryllium windows and the air paths I had in the oscillation camera. My first lysozyme crystal mounted in a glass capillary was immediately disappointing. As I developed the exposed X-ray film in the dark room, I realized my mistake of not allowing for the absorption of the glass walls of the capillary and so I simply cut out a piece of Mylar, placed a crystal in the centre along with the usual blob of mother liquor and rolled the Mylar into a cylinder. I inserted a short length of liquid at one end and sealed it up with melted wax. The SRS 7.2 X-ray diffraction pattern now was readily visible. I did not think to publish this at the time but instead, much later, indeed about 15 years later, I was working on a new protein with expected disulphides that would allow me to increase their f'' anomalous signal considerably by working at such a wavelength. This we did: (Mr Michele Cianci a research student and Dr Andrzej Olczak, an EC funded post-doc, both with me at Manchester University). The SRS 7.2 was un-mothballed, by our collaborator at the SRS Dr Pierre Rizkallah, which it had become in the later 1990s, and we happily played for many days optimizing the set-up for use at 2 Å wavelength. This increased the sulphur f'' significantly, but most notably set us nicely to increase the f'' for the xenon LI absorption edge (at 2.27 Å). We also had the use of SRS 9.5, which had a CCD, and SRS 9.6 also with a CCD. Thus, the crystal structure of apocrustacyanin A1, with crystals grown by Prof. Naomi Chayen, was solved (56). I also wrote a short note featuring the original lysozyme 2.6 Å wavelength diffraction pattern for the new conference series launched by Japanese colleagues: the International Symposium on Diffraction Structural Biology (ISDSB) (57).

4.6.2 SRS Station 9.6: the first super conducting wiggler (SCW) protein crystallography instrument and electronic area detector initiatives

With SRS 7.2 well underway, with firm user community support, a new opportunity arose to expand the technical specification of what could be made available to SRS PX users with the advent of the SRS superconducting (5 T) wiggler (Figure 4.8). This wiggler had a critical wavelength of emission of 0.9 Å, and thus opened up this portion of the X-ray range for use. In addition, it had a higher intensity than even at the SRS 7.2 favoured wavelengths range such as 1.3–2 Å due to the simple fall-off of the SRS spectral curve for a bending magnet field of 1.2 T, with its critical wavelength of 4 Å. As I analysed the new station layout and the fact that this wiggler magnet could provide 60 mrad in total angular width of beam, rather than the 28 mrad of the bending magnet, not only could we have 5 mrad for our new protein crystallography instrument (even if we were not able to secure the prized end-of-beamline position), we could still have a 'straight through beam setting for the 2θ arm". This would allow a white beam to pass through to the sample. At this time, 1984, the Cornell High Energy Synchrotron Source (CHESS) group of Keith Moffat published their seminal paper in *Science* of advocating Laue diffraction for rapid data collection in protein crystallography for time-resolved structural studies in the crystal (58). Not only did this resonate with the vision of David Phillips of harnessing the flow cell of Hal Wyckoff (Yale University; published 1967 (59)), it also became clear later that Louise Johnson and Janos Hajdu with colleagues would become very interested in this option at SRS. Phil Moore was again the mechanical engineer. He and I could build on our design for SRS 7.2 for the 2θ arm, but this time also allow it to be set to this straight-through position. With the computer-aided design system drawings, it became clear that the film oscillation camera, and soon to be added Enraf Nonius (Fast Area Sensitive Television (FAST)) 'TV diffractometer', would collide with the lead shield wall that separated our hutch from the central beamline carrying beam into the outer hall of the SRS. Clearly this was going to be a difficult challenge, but Phil could see how keen I was and he checked with the SRS Radiation Safety Group. They concluded that a small piece cut out of the lead shielding at the 'savage point of interaction' with the lead shield wall would not be a radiation hazard when we were in the SRS 9.6 hutch.

On the detector provision for SRS 9.6, I wrote up the strategy for electronic area detection (60), heavily influenced by Uli Arndt's thinking (61), but extending the analysis from 8 keV only so as to span our SRS interests from 0.5 to 3 Å wavelength. I wrote up the SRS 9.6 design and first applications in ref. (62).

A curious feature of the SRS wiggler installation and firing up to 5 T was that the vertical divergence on SRS 7.2 doubled in value. We first noticed this as a simple decrease in the ion chamber current reading of the SRS 7.2 X-ray beam intensity onto the protein crystal sample. Andrew Thompson, who had joined in the SRS protein crystallography effort, and I investigated this in an SRS Outer Hall beamline 7 test hutch where a simple green paper test showed that for the wiggler off the line 7, vertical beam divergence was 2 times smaller than for wiggler on. I reported our evidence to the machine group, Mike Poole, who made their own checks on their SR beam monitor. It was announced that the wiggler would indeed need a slight planarity adjustment as the magnetic field was 0.75 degrees from horizontal. This was duly corrected, and we got our beam intensity back on SRS 7.2!

The SRS 9.6 commissioning team by now included Andrew Thompson and also Dr Miroslav Papiz who joined me as a Post Doc Research Assistant to commission and implement the FAST TV diffractometer that we purchased from Enraf-Nonius. This effort commenced in 1984. It was a very busy time. The research and development and the expanded user programme (*62*) on SRS 9.6 broke new ground in various research areas. These included:

- Optimized anomalous scattering at the L absorption elements such as Pt, Au and Hg, i.e. the common heavy-atom derivatizing elements (*62, 63*).
- Assessing how far small protein crystals might be pushed, in fact as small as 20 μm was tested in a NATO-funded collaboration I had with Keith Hodgson and Britt Hedman (*64*).
- Broad bandpass Laue diffraction tested with pea lectin crystals (a gift from Howard Einspahr from the SRS 7.2 collaboration I referred to earlier) (*55*), and which led to a whole new software package for evaluating such diffraction patterns (*65*) in a collaboration with Daresbury colleagues Pella Machin and Mike Elder.
- By use of the pre-monochromator slits being made narrower, an unusual large lattice effect crystalline disorder was resolved in crystals of the enzyme rubisco (*66*).
- The extensive and pioneering studies of catalysis in the crystal, and including prompt take-up of Laue diffraction studies, by Oxford Molecular Biophysics (*67, 68*).
- Finally, very high profile (reaching even the BBC 9 o'clock news for the FMDV study) was the large unit cells user programme, which included David Stuart with Porton Down on Foot and Mouth Disease Virus (FMDV) (*69*), Steve Harrison and Bob Liddington from Harvard University with their SV 40 virus (*70*) and Ada Yonath for ribosome crystallography (*71*).

It is important to point out also that the NSLS in Brookhaven, United States, generally labelled a second-generation SR source, in fact had a superior brightness compared with the SRS (Prof. Michael Hart, personal communication, Figure 4.11) and allowed pioneering ribosome crystallography studies to move forward (using a CCD device and with improved crystals by that time) (*72*).

The above SRS 9.6 studies all used photographic film. The Enraf-Nonius FAST TV diffractometer proved a difficult commissioning challenge. Miroslav and I took time in the single bunch low current mode of SRS, as well as portions of the regular multi-bunch SRS operating mode, to undertake tests as well as get familiar with the software that came with the device for the diffraction

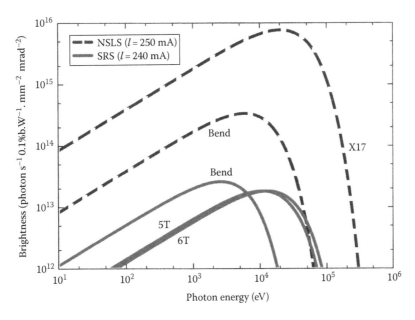

Figure 4.11 Comparison of the spectral brightness of the SRS and the NSLS (Figure kindly supplied by Prof. Michael Hart with permission).

data image processing. We noticed that the noise floor from image to image varied substantially; this led to a Peltier device cooler being added by the company, but this took time to report to the company our findings and accept their technical solution. Since the device did not simply progress into user operation, the SRS senior management started taking a close interest. After all, the device would reduce our substantial photographic film budget: (large quantities of CEA X-ray films were regularly purchased from Sweden as well as dark room supplies of developer and fixer). In addition I had planned this as a major speed-up of the 'measurements to processed data' pipeline; indeed a contract with Enraf-Nonius for this software described reaching 600 processed reflection intensities a second from the software on a reasonably powerful computer (also controlling the apparatus).

In spite of these commissioning challenges, clearly more than teething troubles, the FAST broke new ground: e.g. working with GLAXO (Dr Alan Wonacott), it was used in their drug discovery protein crystallography programme using SRS 9.6 FAST data *(73)*. Secondly, a spin-off application was for a small molecule microcrystals research programme funded by SERC to Dr Marjorie Harding, then at Liverpool University, and which employed a Post-Doctoral Research Assistant, Dr Pierre Rizkallah. This led eventually to a major new application area for SRS of chemical crystallography and ultimately with dedicated beamline instruments SRS 9.8 and later SRS 16.2. An example crystal structure from the FAST on SRS 9.6 for a very small crystal of piperazine silicate is described in ref. *(74)*; this paper includes a calculation allowing a comparison of the scattering efficiencies of relevant small crystal project challenges of the time. A summary and review of chemical crystallography with SRS, and SR in general, was presented by Dr Marjorie Harding up to 1995 *(75)* and by Prof. Bill Clegg on the SRS 9.8 and 16.2 utilization years *(76)*.

An overall survey of the timeline of different detectors used at the SRS for protein crystallography is shown in Table 4.1: this table includes reference to SRS Beamline 14 for protein crystallography led by Dr Colin Nave *(77)*. The overall view of the SRS during this time period up to 1985 is summarized by the SRS Director of the time, Dr Jerry Thompson in ref. *(78)*.

4.6.3 *The SRS High Brightness Lattice and SRS 9.5 for rapidly tuneable*
protein crystallography and point-focussed Laue crystallography

In 1985, the SRS Machine Group brought about the 'High Brightness Lattice' (HBL) project *(79)*, whereby the addition of new focussing magnets in the electron storage ring gave a reduction of mainly the horizontal source size from 14 mm to ~2 mm, and a slight improvement of the vertical source size, which was already only ~0.4 mm. Thus, with no modification of the SRS 7.2 or SRS 9.6 beamline optics, the focal spots improved from 1.4 mm to 0.2 mm or so (FWHM) values. Since the typical protein crystal sample size was 0.3 mm or less, this was obviously a considerable improvement of X-ray beam intensity intercepted by the sample and data exposure times could reduce accordingly. The rate-limiting step was still in effect the film processing in the dark room, which was even more of a bottleneck. The FAST diffractometer software proved to be rather cumbersome, since a crystal orientation matrix had to be calculated and adapted if crystal slippage of the crystal in the capillary occurred. This was still in the days before the freezing technique was pioneered for ribosome crystallography by Ada Yonath and Hakon Hope *(80)*. This technique had already been demonstrated by Rossmann and Haas much earlier with lactate dehydrogenase *(81)*, and it became more widely adopted and optimized: (see the review and initiatives of Prof. Elspeth Garman *(82)*). The capillary mounting technique did quite often lead to sample slippage during data collection.

The HBL however created a different and new opportunity, which was to harness a different style of beamline optics from SRS 7.2/9.6, namely, a double-crystal monochromator with a toroidal mirror for the point-focussing of the beam. Thus, this X-ray spectroscopy style monochromator allowed for rapid wavelength tuning for ease of multiple wavelength anomalous scattering measurements around an elemental absorption edge. The water-cooled channel cut monochromator was provided by Prof. Michael Hart to his own design. There was a spare beamline slot for development on the SRS line 9 wiggler. I obtained funding support from the Swedish Natural Sciences Research Council

Table 4.1 A timeline of the detector's provision at the SRS protein crystallography stations. X-ray film was the mainstay of research in the 1980s, noted for the length of time one used to spend running between station and dark room, and the equally onerous task of digitizing the precious diffraction patterns. The FAST area detector drew the user's attention to the advantages of electronic detectors. Image plates quickly spread around protein crystallography (PX) facilities across the world, including the SRS, through the 1990s, until CCD detectors became available

	82	83	84	85	86	87	88	89	90	91	92	93	94	95	96	97	98	99	00	01	02	03	04	05	06	07	08
7.2	Film													18	MAR 30			MAR 345				Test			Decomm		
9.6	Not built			Film		Film + FAST					R-axis	MAR 30				ADSC Q4	188 × 188 mm 2400 × 2400 pixels						Non-PX		Test		
9.5	Not built							Film			MAR 18 cm			MAR 30 cm					MAR 165 mm CCD								
14.1	Not built																			ADSC Q4							Test
14.2	Not built																			ADSC Q4							Test
10.1	Not built																					165 CCD		MAR 225 mm CCD			

Note: Prepared by Dr Pierre Rizkallah and reproduced with permission. The Beamline 14 initiative was led by Dr Colin Nave (77).

and who provided a Post-Doctoral Research Assistant, Dr Ron Brammer, who came from Sweden to Daresbury. Ron led the X-ray optical ray tracing details, and this work and results were published in *Nuclear Instruments and Methods* in 1988 (*83*). Andrew Thompson was Instrument Scientist for SRS 9.5. This beamline opened up Multiple-wavelength Anomalous Dispersion (MAD) measurements in a routine way from protein or nucleic acid crystals and first results were obtained (*84*).

4.6.4 *The ESRP and the ESRF planning*

In Europe, discussions on an ESRF third-generation machine had been going on since the mid-1970s to late 1970s (*85–89*). My first chance to join in such discussions was at the Life Sciences Workshop held in Oxford 25–27 February 1979 (*87*). I was able to contribute the practical experience of my DPhil project having recorded X-ray oscillation film camera exposures of single crystals on 6-phosphogluconate dehydrogenase on a rotating anode GX6 at the Laboratory of Molecular Biophysics in Oxford University, and then also at the LURE synchrotron Paris. The exposure times were, respectively, 10 h to 2.3 Å diffraction resolution and 24 min to 2 Å resolution. Therefore, this extrapolated to 15s for an equivalent exposure at ESRF. I further commented on the benefits of ESRF for small crystals (*90a*) as follows:

> The effect of reducing the crystal dimensions on increasing the exposure time is large at the national sources but (will be) small on the ESRF where the beam cross-section can be reduced to accommodate the size of the crystal.... In principle then a 0.1 × 0.1 × 0.1 mm³ crystal would take 7 hours to expose on a present day SR source but only 35 seconds on the ESRF.... Theoretically it should be possible to get information in periods of hours from very small crystals, in the region of 10–20 microns per side.

The other sections in the report on protein crystallography at the proposed ESRF, besides the one just referred to, named 'Amplitude measurements', were entitled: Anomalous scattering phase measurements, Kinetic experiments on crystals and Practical considerations at the ESRF. Other practical protein crystallography data highlighted were high-resolution data for glycogen phosphorylase b, measured at LURE by Enrico Stura and Keith Wilson, including diffuse scattering (*90b*). The chemical crystallography applications were referred to in the section entitled 'structures' with long range three-dimensional order' with sub-headings: Determination of charge density, Crystal structure determinations (including single-crystal work and powder work), Magnetic materials and Order in disordered structures. The section 'Kinetic experiments on crystals' was suitably cautious 'since the most primitive measurements have only just begun', but 'it may be possible to obtain a diffraction diagram (from a protein crystal) in less than one second'. The use of the Laue method however was not anticipated in this report, neither that the focussed white nor pink beam from an ESRF wiggler-undulator (wundulator) would allow single-bunch exposures from a protein crystal, i.e. with a sub-nanosecond single pulse being sufficient to obtain a protein crystal Laue diffraction pattern (see below).

A report on progress was presented by Buras (*91*), which included detailed comparisons of the available and planned SR sources' spectral brightness (Figure 4.12). Likewise in the United States, Stanford Synchrotron Radiation Lab (SSRL) was hosting New Rings workshops on how to harness the PEP storage ring as an SR Source and which, like PETRA in Hamburg, had the possibility to realize the ultimate storage ring machine design for SR applications.

Spectral brightness was firmly recognized as a particularly important parameter (at that time often referred to as spectral brilliance). The ESRF design was set to outperform any previous or planned source, its closest competitor in the spectral brilliance graphs presented being the NSLS (*91*), Figure 4.12.

With Keith Hodgson and Britt Hedman (SSRL), I obtained a NATO-funded collaborative research grant to try and evaluate these new SR X-ray sources: 'ultra-high-flux-brightness rings'. By using the focussed SRS wiggler white beam for Laue diffraction, the intensity onto the sample was a mimic of the ESRF monochromatic undulator X-ray beam intensities, typically envisaged

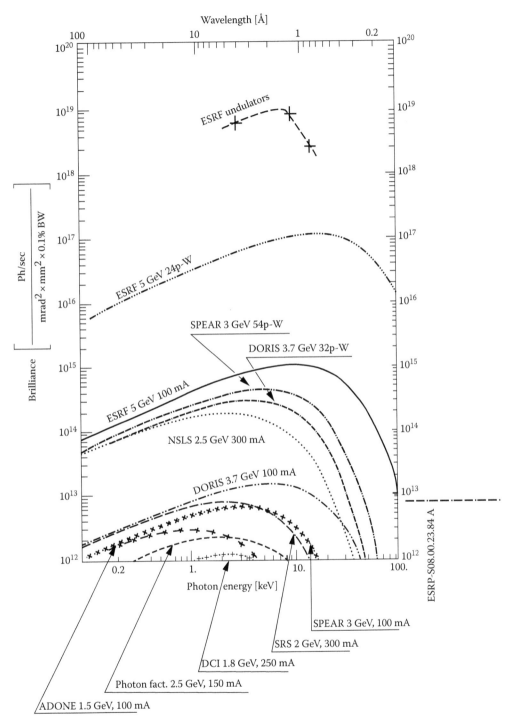

Figure 4.12 Spectral distribution of brilliance for the ESRF bending magnet, 24-pole wiggler and several undulators and, for comparison, the brilliance from, bending magnets on ADONE (Frascati), DCI (Orsay), DORIS (Hamburg), NSLS (Brookhaven), Photon Factory (Tsukuba), SPEAR (Stanford), SRS (Daresbury) and the 32-pole wiggler on DORIS and the 54-pole wiggler on SPEAR. The brilliances for the MPWSs are on axis. From ref. (*91*) Courtesy of Brookhaven National Laboratory.

to be around 10^{14} photons/s/mm² onto the sample. We conducted two main evaluations: one evaluation was 'could we measure successfully X-ray diffraction from protein microcrystals?' and the second was 'what was the chemical nature of X-ray radiation damage after prolonged dose which would occur as the sample size was made progressively smaller?' The first evaluation we published in *PNAS* (*64*). Since clean diffraction data were obtained from a 20 μm crystal, this immediately meant that finer sample alignment on the diffractometer, which should be 10 times better than sample size, would certainly be needed (i.e. 2 μm) and which was much tighter than the typical manufacturer's diffractometer 'sphere of confusion', which was then around 25 μm.

The second evaluation with Keith Hodgson and Britt Hedman built on a suggestion of Prof. Greg Petsko of Brandeis University that a marker of X-ray dose damage to a protein could be the splitting of disulphide bridges. Thus, we examined this with monochromatic data measured both at Stanford and at Daresbury and revealed for the first time that this had in fact occurred, with negative density where the two sulphur atoms had been and positive density where they had moved to, i.e. they had moved apart (*92*).

Dr Roger Fourme, who was leading the LURE, Paris synchrotron protein crystallography programme, and I produced a report for the European Synchrotron Radiation Project (ESRP), based at CERN in Geneva. The ESRP was the precursor stage before the ESRF had been agreed to be funded (and therefore well before the site of the ESRF was decided upon by the partner countries). This report (*93*) offered projections on the utilization of these totally new, high level, monochromatic undulator intensities. The report included simulations of beam heating of the sample and a risk management strategy involving a novel crystal sample mount on a copper fibre with a heat-conducting glue or grease to attach it.

These sample beam-heating calculations were of a simple adiabatic modelling type, introduced by Kam previously (*94*). Later simulations involving detailed isothermal interactions were modelled (*95*), and were followed up by actual heating measurements by Snell et al. (*96*). Again microcrystals were advised to be a territory of user science for SR where there was a risk of adverse beam-heating effects. Snell et al. concluded as follows: 'However, microcrystals require more careful considerations. The integrated intensity of the diffraction peaks is proportional to the product of flux intercepted by the crystal, crystal thickness and exposure time per frame. If the experimenter chooses to increase the flux on the sample instead of increasing the exposure time to make up for the smaller sample thickness in order to achieve the same intensity of the recorded diffraction peaks, they may steer into dangerous territory. The same caveat applies to radiation damage studies where maximum fluxes are used instead of just maximum flux densities. A flux of 6×10^{12} photons s⁻¹ intercepted by the sample, i.e. four times the 'typical' flux used for the extrapolation, would result in a predicted temperature rise of 20 K (24 K for a 0.025 mm-diameter sample). This, indeed, would bring the sample close to the onset of enhanced free-radical mobility with associated consequences for radiation damage'.

These pessimistic prognostications for the ESRF applications territory of biological microcrystals were countered by the prudent experimental tests on SRS 9.6 with the focussed white beam that led us to publish Hedman et al. (*64*). These kept the risks for this area of application at a level that did not impede our proposal for microfocus applications of ESRF including for microcrystals. Of course, Hedman et al. (*64*) had used gramicidin crystals, where gramicidin, although biological, is a rather small protein, indeed some would say only a polypeptide of 3 kDa molecular weight. That said Hedman et al. (*64*) also included tests with the focussed white beam on normal sized haemoglobin crystals, and which still produced clean Laue diffraction patterns, although suffering radiation damage after several exposures. ESRF did indeed set up a microfocus beamline facility in the first phase of instruments and which were advertised under an ESRF Newsletter article by the Science Director of the time Prof. Andrew Miller (*97*) that macromolecular micro-crystallography could expect 25% of the beamline allocations on it. This application area grew substantially under the excellent leadership of Dr Christian Riekel (*98*) and ESRF became a global leader in the area for many years, and indeed, other facilities were subsequently engaged in 'catch-up'. The Advanced Photon Source at Argonne National Laboratory (APS) was quick to respond with more than one of the ESRF EMBL microdiffractometers (*99*) being purchased and installed on beamlines, such as NE CAT and GM CAT.

The overall point about the experimental tests, simulations and calculations described above was to prove sufficient to carry the community usage plans forward in conjunction with the ESRF Facility. I came across a most interesting reference in writing up this Lecture. It is by Holton and Frankel (*100*) who assessed the question of the minimum crystal size for a complete diffraction data set. This study mainly worked within an ideal limit of zero background. Their abstract states that a crystal of lysozyme of 1.2 μm could yield 2 Å diffraction data, and if photoelectron escape models are included (*101*), then even a sample as small as 0.34 μm could be used. Adding that with the current 'detection limit' whereby '100 photons/*hkl* are needed (after data merging of likely symmetry related or multi-measured spots) to attain a signal-to-noise ratio of 2 (means that), a lysozyme crystal will have to be 8.3 microns in diameter for 2 Å data, and (for) 3.5 Å data for a 10 MDa case would require 43 micron crystals, limiting the usefulness of X-ray beams smaller than this'.

The ESRF Foundation Phase Report (known as The Red Book) (*41*) comprised ~1000 pages of machine and beamline descriptions; the undulator spectral emissions, for example, at that time are shown in Figure 4.9. The Working Group for macromolecular crystallography proposed the provision of at least a bending magnet for MAD use, a MPW for monochromatic and white-beam Laue work, a monochromatic undulator beam for large unit cell and small crystals and, more speculatively, a beamline for providing very high photon energies to access high atomic number K edges (such as Pt, Au and Hg) for optimized anomalous scattering. This last request was never serviced, but a generic high energy beamline was established and some tests ensued much later. A very broad instruments' provision for macromolecular crystallography now exists at the ESRF with these various sources led by Dr Sean McSweeney.

The first ESRF Science Advisory Committee (SAC) decided to establish generic beamlines emphasizing specific beam properties such as 'High flux', 'white beam', etc. Whilst it was an expeditious way to serve many research communities, it was a setback to expectations to access substantial portions of beamtime for macromolecular crystallography. Roger Fourme and I complained about this in writing to the ESRF SAC. An outcome of this was that macromolecular crystallography was represented at the second ESRF SAC; Prof. Jens Als Nielsen, Copenhagen Physics Department, was elected as Chair, and I was elected Vice-Chairman. As Vice-Chairman this meant I represented the SAC at the ESRF Machine Advisory Committee (MAC), as the SAC Chair Jens had to attend the ESRF Council.

Thus, I had at least four meetings per year in Grenoble, two SAC and two MAC meetings. The MAC was chaired by Dr Jerry Thompson of Daresbury Laboratory, the SRS Director. The number of meetings in Grenoble escalated further as the concrete floor to the ESRF Experimental Hall was laid incorrectly by the contractor (who, much later, had to pay damages costs to the ESRF). The Concrete Working Party met several times, under increasingly vitriolic argument both within the in-house Directors (the Director General and the two ESRF Science Directors) and within the ESRF SAC. By this time, Prof. Jen Als Nielsen had joined ESRF and had to vacate the ESRF SAC Chair; as Vice-Chair, I assumed the role of Chair. This meant I chaired the joint MAC with SAC meeting which brought the issue to a head with all competing factions and personalities present. It proved to be the most difficult meeting I have ever chaired. Two extreme views were expressed. The Director General Ruprecht Haensel emphasized a civil engineering solution known as 'shallow grouting', and which involved squirting liquid grouting compound between cracks in the floor slabs every 12 months; this seemed immediately unsatisfactory as it was meant to be a clean scientific working environment which would be forced to be a 'building site' every 12 months. The contrary view championed by Prof. Michael Hart was that the concrete floor should be dug up and taken away by the contractor who should then lay a new floor; this option seemed also unworkable as not least the sheer quantity of concrete to be dug up and lorried away was very large. Prof. Massimo Altarelli, from within the ESRF Management, proposed that the civil engineering consultants continued to hunt for a more long-lasting solution to the problem. This option which faced the joint MAC/SAC meeting had the danger of looking like procrastination if we took it, but in spite of that disadvantage, it was the only realistic thing to commend to ESRF Council, who had the ultimate authority over such decisions. As ESRF SAC Chair, I attended the next ESRF Council. Prof. Jules

Horowitz was in the Chair and was greatly experienced in all manner of large building projects. The civil engineers did indeed by that time offer the 'deep grouting' solution to the problem and for which they estimated that the floor would be rendered stable and sufficiently flat for up to 20 years. This was received with great relief by all participants, not least as engineers' safety factors in any project would likely be such that it could be good up to twice that length of time! Whilst the whole problem was lamentable that it had even occurred in the first place, it was a profoundly satisfying business to be resolved. Of course, most importantly, it was a great thing to be involved with such a supranational pan-European project.

The ESRF put European scientists involved with SR and its applications in a global leadership position. Approximately 2 years later, the United States followed with the 7 GeV APS and not long after the 8 GeV SPRing-8 Japanese project was approved. Innovations were not only the domain of ESRF in SR science and its evolution. The APS introduced the top-up machine operating mode so that beam currents were essentially constant: this had two benefits, firstly, the integrated flux was obviously larger per unit time and, secondly, beamline optics' stability under a constant thermal load was much better than a decreasing beam current situation. Also the 7 and 8 GeV machine energies allowed a wider tuning range as undulator magnet gaps were varied, in effect giving a continuous sweep of X-ray energies from a single undulator emission fundamental or harmonic.

From the ESRF, APS and SPRing-8, the necessary experience was gained in both the machine and science domains to generate the designs for the new 3 GeV SR national machines such as the UK's DLS. Using narrower undulator gaps, pioneered at the NSLS (*42*), the lower machine energy of around 3 GeV still allows undulator emission into the most popular portions of the X-ray region of the emission spectrum.

4.6.5 *ESRF BM14: the first macromolecular crystallography instrument on a third-generation SR source*

Following the details of the SR source evolution described in the previous section, it is worth mentioning some of the details of the first macromolecular crystallography instrument on a third-generation SR source. The ESRF BM14 design was based on the SRS 9.5 design. It was a development led by Andrew Thompson who had joined ESRF from SRS. A weakness of SRS 9.5 was that the vertical divergence angle had to be controlled by vertical slits and which led to a weaker beam intensity than users had come to expect on SRS 9.6 in particular. The ESRF BM14 could instead accommodate a collimating pre-mirror, pioneered at NSLS (*102*), instead of slits; and thus, a gain in intensity of ESRF BM14 over SRS 9.5 of a factor of about 20 was immediately apparent.

At this point, since this lecture was also a Teaching Plenary for the BCA Biological Structures Group, it is appropriate to explain the key facts of phasing in protein crystallography. Using isomorphous or anomalous differences at a single wavelength leads to two possible choices of the phase angle for a reflection, shown in Figure 4.13 *via* the two vector triangles which are compatible with the measurements. A third measurement is needed, for example, from a second isomorphous heavy atom derivative, to realize a unique phase value determination (not shown), summarized in (*103*). If the crystal sample contains a metal atom, then data measured at two wavelengths are adequate with the anomalous difference to yield three measurements and a unique phase determination. Since nearly all proteins contain methionine, which contains sulphur, it is possible to harness selenomethionine, where the selenium has identical chemistry, being in the same chemical group of the periodic table. Thus, the rather inaccessible sulphur K edge of 5 Å wavelength is much easier for the selenium K edge of 0.98 Å. The selenomethionine approach as a general phasing vehicle for proteins was the brainchild of Wayne Hendrickson (*104*). In bringing to the fore SRS 9.5 and ESRF BM 14, along with demonstrating the utility for such measurements of the ESRF Image Intensifier electronic area detector, we collaborated with Dr Alfons Haedener of Basle University who had selenomethionine hydroxy methylbilane synthase crystals. A successful MAD structure determination on both instruments of this enzyme in its active, i.e. chemically reduced form, resulted (*105*).

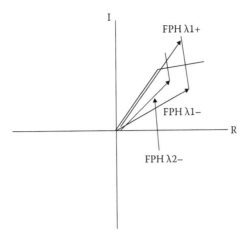

Figure 4.13 The unique solution for each reflection phase stems from three unique measurements from a minimum of two wavelengths at an absorption edge. Illustrated here are both the reflection and its Friedel mate at λ1 and the Friedel mate only at λ2. See also Ref. 50 Figures 52 and 53.

4.6.6 *SR Laue crystallography at SRS and ESRF*

The development of the Laue method for quantitative crystal structure analysis was an important new thread to the development of the field of SR crystallography. This development followed the initiative of Moffat and colleagues at CHESS Cornell (*58*). On seeing this paper, I immediately felt it was an exciting development in which we at Daresbury should join. I had already made sure that SRS 9.6 could take a straight-through beam into the crystallography camera or FAST diffractometer, as mentioned above. This could be either a white beam or a monochromatic beam from a double-crystal monochromator. I recorded the broad-bandpass Laue diffraction patterns from a pea lectin crystal during single-bunch (i.e. normally a non-user) mode during early 1984. These Laue patterns were of a very high quality of signal-to-noise ratio and immediately looked promising for quantitative analysis. By chance I was looking at the proofs of my *Reports on Progress in Physics* review article and asked if I might be allowed to show a figure or two of these protein crystal, broad wavelength bandpass, Laue diffraction patterns, which was granted by the IOP Publishers (*50*). These Laue patterns were of a different kind to those published by Moffat, Szebenyi and Bilderback, who had carefully explained the need for a wavelength bandpass of 0.2 so as to avoid the overlapping orders problem of Laue diffraction.

At Daresbury, colleagues Dr Mike Elder and Dr Pella Machin ran the SERC Microdensitometry Service for digitizing Weissenberg diffraction films and these data were passed back to the crystallographer for crystal structure solution. They were immediately captivated by the high quality in terms of signal to background of the pea lectin Laue diffraction patterns. Mike said he would write a Laue diffraction spot prediction program and asked me for the experimental parameters of the white SRS 9.6 X-ray beam and the pea lectin unit cell and space group details. The beam spectral parameters were $\lambda_{min} = 0.2$ Å and $\lambda_{max} = 2.6$ Å. The short wavelength at that time of the SRS 9.6 instrument was as yet unmodified by the SRS 9.6 reflecting mirror, which was installed later. The long wavelength cut-off was approximate and set by the beryllium window transmission. I told Mike about the 'overlapping orders' challenge: see for example Bragg (*106*), from which I quote:

> X-ray analysis started with the Laue photograph. Although this method was developed further by Wyckoff in America (*107*), with a considerable measure of success, it never came into general use. It is too hard to attach a quantitative significance to the intensity of the spots, which are due to the superposition of diffracted beams of several orders selected from a range of 'white' radiation. The (monochromatic) spectrometer proved to be of much greater analytical power because it measured accurately a diffracted beam of monochromatic radiation of definite order, and the first crystal structures were solved by it.

That said, Bragg's analysis of sodium chloride and related crystals (*108*) is a *tour de force* of Laue diffraction photos recorded by him in Cambridge, and quantitative analysis along with complementary monochromatic measurements, the latter measured with his father W.H. Bragg in Leeds.

Mike Elder looked for the multiplicity effect and displayed a histogram of how many predicted pea lectin Laue spots were singlets, doublets, triplets, etc. Of special interest for him was the particular type of computer workstation for which he would write the program. This was a PERQ computer which had been bought in a bulk buy by SERC as a special initiative of high-performance (at the time) distributed computing workstations. It also had a high-quality display screen, albeit black and white, well suited to displaying the predicted Laue spot patterns, comprising up to about 10,000 reflections, for the particular crystal and beamline parameters. This program was very promptly written in a few days by Mike Elder. It immediately showed that the predominant population of spots, by a long way, were singlets. Further analysis proved that the low-resolution diffraction reflections were nearly always members of Laue spots that were multiplets. The initial difficulty in extracting the intensities of Laue spots containing several 'Bragg reflections' led to our test crystal structure analyses initially being based around the population of Laue diffraction spots that were singlets. These singlet Laue spot intensities still needed wavelength normalization to bring them all onto a common intensity scale (*109*). Another Daresbury computer programmer, John Campbell, focussed on this development and wrote a computer program called 'LAUENORM' (described in (*65*)).

The pea lectin Laue diffraction patterns were duly processed and the details of this and the software were published in ref. (*65*). The theoretical properties of the multiplicity distribution were studied in detail as well by Durward Cruickshank, Keith Moffat and myself (*110, 111*).

The Oxford University team was also quick to see the potential of this synchrotron Laue method development for application to time-resolved crystallography studies of their phosphorylase enzyme, thus extending their results in *EMBO Journal* (*67*) on 'Catalysis in the crystal: synchrotron radiation studies with glycogen phosphorylase b' and publishing a study: Hajdu et al. (*68*). Clearly these results and developments affirmed the excitement of Laue diffraction identified by Moffat et al. (*58*).

We also undertook a wide range of validation experiments to establish both the credentials of the Laue method and our own, that is, the 'Daresbury Laue software package'. In my Lecture, I illustrated this major R&D programme with a validation evaluation on an amylase enzyme crystal, kindly provided by Zygmunt Derewenda (*112*). This study included deconvoluted doublet spots into two reflection intensity components, which yielded clearly the mercury atom site from the difference Fourier map. Furthermore, at the time, there were critical remarks about the difficulty of including a focussing mirror, in terms of focussed beam instabilities at the sample. I remarked 'Figure 4.3(*112*) provides the "acid test" for data quality with a mirror and shows difference Fourier maps for Laue and monochromatic data. Figure 4.4 (*112*) used phases derived from Laue Hg amylase data; the three sections indicate that the anomalous differences are very significant when measured in Laue geometry with a mirror'. These results I presented in a lecture at the International Synchrotron Radiation Instrumentation Conference in Tsukuba, Japan, (1988) and published in *Reviews of Scientific Instruments* in the Conference Proceedings (*112*).

The ESRF 'white beam' beamline ID09 aimed to serve various research communities, such as time-resolved Laue diffraction and high-pressure experiments. Dr Michael Wulff was the Beamline Leader for ESRF ID09. He came to SRS Daresbury in the early 1990s to see SRS 9.5 for himself and how we had achieved a point-focussed Laue beam using a toroidal reflecting mirror. Andy Thompson was the SRS 9.5 Instrument Scientist and we measured example Laue patterns. The mechanical engineer for this station was Neville Harris.

Michael, with ESRF ID09, brought about such intense point-focussed white beams, initially broad bandpass from an MPW but eventually preferring narrow bandpass from a tapered undulator, that a single bunch of electrons in the ESRF ring produced a measurable Laue diffraction pattern from a protein crystal. The notion of taking such a single bunch had been advanced by Prof. Keith Moffat and by which the time resolution for studying structural changes in a crystal became the time width of the electron bunch, namely, 100 picoseconds or so. In the case of a stroboscopically cooperating crystal sample/molecular system, then enough of these pulses could be accumulated to achieve reasonable measuring statistics per time point, after a given light-driven stimulus. In my Lecture, I showed a molecular movie, the study by the Moffat team with Michael Wulff on carbon monoxy myoglobin, in which a light-stimulated carbon monoxide molecule is driven off the haem group, and where it wanders is studied in successive time-lapsed movie frames (*113*). The clock at bottom left of the movie ticks away throughout until a time of 3.16 μs has elapsed.

As we concluded the Daresbury Laue software development, and all the associated analyses procedures, for photographic films, new detector opportunities arose. ESRF ID 09 had an image intensifier TV detector developed by J.P. Moy of the ESRF Detector Group (*114*). This was both more sensitive to X-rays, over a wide wavelength range, than film and its ease of read-out made a dramatic improvement to both the ease of conducting such experiments but also increased substantially the number of Laue diffraction exposures per protein crystal. On moving to the University of Manchester as Professor of Structural Chemistry, I was able to combine my interest in methods developments at the synchrotron for crystallography with my own structural studies research programme. One such study was with Dr Alfonse Haedener of Basle University on the enzyme hydroxymethylbilane synthase. Thus, a mutant of the enzyme proved viable for Laue diffraction, with the highly X-ray sensitive Moy detector, and we could measure a large number of Laue patterns leading to high-quality electron density maps (*115*). We also measured a sequence of time-resolved electron density maps (*115*) as the enzyme crystal was fed substrate *via* a flow cell (*59*), the device first shown to me by David Phillips at my DPhil interview in 1974!

4.6.7 *Synergies of Synchrotron and Neutron Laue macromolecular crystallography: initiatives at the Institut Laue Langevin*

In the early 1990s, I was contacted by Drs Clive Wilkinson of the Grenoble Outstation of the EMBL and Mogens Lehman of the Institut Laue Langevin (ILL) about the possibility of the ILL introducing the neutron Laue method for biological and chemical crystallography with neutrons. The idea was that neutron fluxes were low compared with X-ray fluxes, and by harnessing a wide spectrum of emitted neutron wavelengths, this would open up a range of new and more challenging projects for crystal structure analysis. Thus, in biological crystallography, protonation states (as deuterium) of ionizable amino acids such as histidine, aspartic acid and glutamic acid, as well as more detailed information on the orientation of water (D_2O) molecules, could be determined on molecules which had been previously out of reach of monochromatic neutron beams, such as higher molecular weight proteins and/or smaller crystals. For chemical crystallography, the interest was usually the determination of the anisotropic vibrations of hydrogen atoms, but again many studies were out of reach of neutron monochromatic methods due largely, in this case, to too small a crystal being available (since the molecular weights were by definition relatively small already). Clive and Mogens had kept a watch of the progress with the synchrotron X-radiation Laue method for quantitative crystal structure analysis and also noted that the Daresbury Laue software (*65*) could be readily adapted to processing neutron Laue diffraction data. They proposed then a cold neutron source at the ILL reactor experimental hall for macromolecular (largely protein) crystallography with a mean wavelength λ ~ 3.5 Å, so as to benefit from the crystal sample scattering efficiency increasing as a function of $λ^2$ and from a cylindrical image plate diffractometer, which would be called LADI (Laue Diffractometer). Secondly, for chemical crystallography, a hot source neutron beam, providing shorter wavelength neutron beams, would allow the naturally higher diffraction resolution needs for data collection to be satisfied. This instrument would be called Very Intense Vertical Axis Laue Diffractometer (VIVALDI). Each apparatus involved a cylindrical neutron-sensitive image plate fully surrounding the crystal. The Daresbury Laue software coordinate prediction algorithm (for a flat detector) needed amendment (for the cylindrical detector), as did the polarization correction formula (*116*). A critical feature was the provision of neutron-sensitive image plate materials by

Dr Nobuo Niimura from the Japanese Atomic Energy Research Institute (JAERI) in Tokai, Japan, which was a most helpful and collaborative gesture that was greatly appreciated here in Europe.

Japan took a special initiative in this area of neutron macromolecular crystallography within a major funded project known as an Organized Research Combination System (ORCS) grant from the Ministry of Education, Culture, Sports, Science and Technology of Japan to Dr Niimura and Dr Mizuno (Agriculture Research Institute, Tsukuba). Within this project, many scientists were involved, including many who took part in several international workshops hosted in Japan. Along with Prof. Eric Westhof from France, I was involved on the ORCS Advisory Committee over the period from 2000 to 2005. Thus, I was able to repay our gratitude, on behalf of Europe and myself, and give our thanks to Dr Niimura and Japan for the gift of the neutron-sensitive imaging plates referred to above.

Clive and I wrote up a jointly authored book chapter for the HERCULES Course students book and which outlined the ideas briefly summarized above (*117*).

By a happy coincidence, at the same time as Clive Wilkinson and Mogens Lehmann contacted me for collaboration in the matter of the synergistic possibilities between synchrotron X-ray and neutron Laue diffraction and software for data processing, I was rummaging around the crystallization laboratory in the School of Chemistry, University of Manchester, and saw a large measuring cylinder with a dialysis bag. This had been set up by my PhD student Dr Susanne Weisgerber. This dialysis bag contained several very large ($\sim 5 \times 2 \times 1$ mm^3) crystals of one of our structural chemistry/biology projects of the time, the plant lectin concanavalin A. This opened up our own user programme on LADI-I.

Initially, our objective was to solve a neutron protein crystal structure including the bound water structure for concanavalin A, especially the waters bound in the saccharide-binding site which are displaced on sugar binding. Of course our 0.94 Å X-ray crystal structure of concanavalin A had revealed, *via* bond length analyses, protonation states of various Asp and Glu amino acid side-chains and even, to our surprise, approximately 10 bound waters where the hydrogens were visible in the final difference Fourier electron density map. We labelled this X-ray crystal structure, *via* the title of our article, 'The structure of concanavalin A and its bound solvent determined with small-molecule accuracy at 0.94 Å resolution' (*118*). In effect then the contribution that a neutron protein crystal structure could make did rather focus on the saccharide site–bound water structure. Could we see their hydrogens too? – which were obviously too mobile to be delivering their X-ray scattering signal to a high enough diffraction resolution. Initially, I assumed that this meant that the hydrogens had to diffract to 0.94 Å to see them in X-ray-derived electron density maps. In fact, commented on by several neutron specialists, the rapid fall-off of the hydrogens' X-ray scattering factor means that their main signal is at a lower resolution relative to the other atoms. Thus, the role of 0.94 Å X-ray resolution is in resolving the (slightly) heavier atom with an appropriately small atomic radius of electron density, and that way the heavier atom does not submerge the detail latent in the hydrogen electron density.

In our first neutron concanavalin A LADI-I experiment, we tried to effect our hydrogen to deuterium exchange, of the H$_2$Os and the protein's exchangeable hydrogens/protons, using a sealed jar with two pots; one pot with a crystal in mother liquor and another with D$_2$O, heavy water, in it. This was left sealed for at least a week before crystal mounting and taking to Grenoble by Dr George Habash, a Post-Doc working with me for many years, for measurements on LADI-I. The crystal diffracted well, but to our surprise, there was basically no bound water visible in the nuclear density map presumably because of poor H$_2$O to D$_2$O exchange and the cancellation of the two hydrogens' signal, being negative (2×-0.374) and of approximately the correct magnitude as that of oxygen (0.58), and the example of a protonated Asp, based on the X-ray crystal structure, was that there was some density where the hydrogen was, but it was still a hydrogen, i.e. with negative nuclear density. The manganese site showed clear, albeit not strong, negative nuclear density, as to be expected for the neutron scattering signature of manganese. This study we published as a companion paper to Deacon et al. (*118*), i.e. Habash et al. (*119*). However, it was clear this study needed repeating but with a better method of H/D exchange. So, with our collaborator Dr Joeseph Kalb Gilboa from The Weizmann Institute, who used repeated dialysis against heavy-water-based mother liquor of another large crystal of concanavalin A, over a period of 4 months, more LADI-I neutron Laue diffraction data were recorded. This time the diffraction was better, to 2.4 Å resolution versus 2.7 Å in the previous case. Now the bound waters were available for detailed study in the nuclear density maps (*120*).

Next, we sought to see more clearly the saccharide-binding site–bound waters *via* a cryo-version of (*120*), which had been made at room temperature. This was taken forward, led by Dean Myles of the Institut Laue Langevin, with our joint research student Matthew Blakeley, working with concanavalin A and a few other proteins. This R&D established cryo-freezing conditions even with such big crystals, indeed the biggest, to our knowledge, attempted to be frozen. LADI-I had a 15 K cryopad cooling-type device and so after plunge freezing, a crystal was transferred to LADI-I, where it was held at 15 K. There was a considerable increase of the number of bound waters discernible (*121*) versus the room temperature neutron study (*120*), which in turn was an improvement versus the X-ray cryo-structure (*118*). The saccharide site–bound waters' nuclear density was improved to define their orientation in a network. It is also interesting to emphasize the strategic importance of this for neutron macromolecular crystallography in general, i.e. with respect to freeze trapping of reaction intermediates that could be inducible in a crystal sample.

There was also a push to raise the molecular weight ceiling of such neutron crystallography techniques. Thereby, we undertook a neutron Laue diffraction experiment on a complex of concanavalin A with methyl α-D-glucopyranoside extensively soaked in D_2O (space group $I2_13$, $a = 167.8$ Å), which resulted in 3.5 Å diffraction data. In our long-standing programme of structural studies of crystalline saccharide complexes of concanavalin A, the unit cell of the cubic $I2_13$ complex of concanavalin A with methyl α-D-glucopyranoside was one of the largest. With its cell edge of 167.8 Å and its asymmetric unit of molecular weight 50 kDa, it represented a nice challenge for the then current neutron diffraction technology. The size of the crystal used in the experiment, although large ($4 \times 3 \times 2$ mm³), was not the largest ever produced for this complex! The degree of spatial overlapping of diffraction spots observed in the Laue experiment, however, suggested that use of larger crystals would be a disadvantage. On the basis of these observations, several technical improvements for macromolecular neutron crystallography were suggested (*122*). These were incorporated, along with others, in the LADI-III apparatus and which subsequently came online at the Institut Laue Langevin reactor replacing LADI-I.

Subsequently, led by a new PhD student, Stu Fisher, we analysed the freely available protonation prediction tools regarding their efficacy (*123*) and assessed the risks of deuteration altering a molecular structure (*124*).

A major review and summary of the field of neutron macromolecular crystallography has been published by Dr Matthew Blakeley (*125*). This includes a comparison of the effectiveness of Laue methods in neutron macromolecular crystallography, which shows how the limits of high molecular weight and of small sample, as well as the speed of measurement, have been significantly improved with the neutron Laue method. The reduced background noise that inevitably comes from using the broader spectral bandpass has so far had only a marginal effect on worsening the diffraction resolution limit achieved. The total elapsed time to make a data set measurement has also reduced significantly, as the evolution of apparatus exploiting technical developments has occurred (Figure 4.14). Briefly, the sequence of apparatus developments at the Institut Laue Langevin had been as follows. The LADI apparatus was an EMBL development. It became known as LADI-I, i.e. first version, as VIVALDI for chemical crystallography was in effect LADI-II. Later the LADI-III replaced LADI-I with improved features, which were that the readout head was on the inner surface of the image plate, yielding an improved DQE (16% for LADI-I and 46% for LADI-III) and was therefore close to a factor of 3 times more sensitive (*126*). The radius was also somewhat larger, thus allowing a larger crystal and subsequent

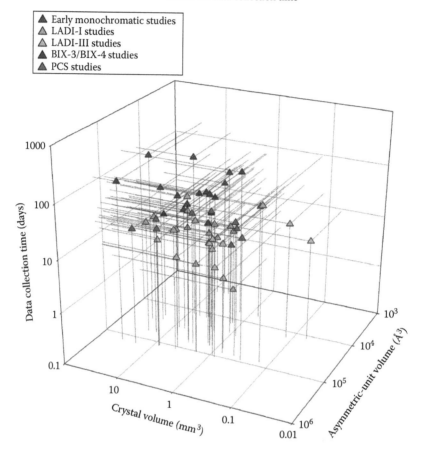

Figure 4.14 (See colour insert.) Three-dimensional scatter plot of the asymmetric unit-cell volume vs. crystal volume vs. data collection time for the various neutron structures solved so far. The new instruments are able to collect data from larger asymmetric unit cells, from smaller crystal volumes and with shorter data collection times. The original necessity to have crystal volumes of several mm³ is no longer the case with data now being collected from perdeuterated protein crystals close to 0.1 mm³. (Includes unpublished results from LADI-III.) From ref. (*125*) with permission of the author and Informa UK.

spot-to-spot spatial resolution to be better. Also, for a given crystal sample size, the unit cell diffraction order-to-order resolution could be improved.

In the spirit of the teaching aspect of the Lonsdale Lecture, some basics of neutron scattering are as follows. The neutron scattering factors of each element show quite different trends compared with the X-ray scattering factors. For neutrons, the scattering cross-section magnitude shows relatively very small variations and with some quite remarkable characteristics, such as deuterium scatters basically the same as carbon. Thus, for neutron macromolecular crystallography, even with modest resolution, neutron diffraction data such as 2.5 Å or better, the full hydrogenation (as deuterium) details can be discerned. The perhaps more famous application of neutrons in biology is of course the contrast variation method where by suitable adjustment of D_2O versus H_2O relative percentages can lead to contrasting in or out a protein or a nucleic acid in a multi-macromolecular complex. Alternatively, the specific labelling with deuterium of molecules in a complex can alter their scattering signature and thus allow their placement to be determined. A third, yet to be applied in earnest application, is the spin polarization property of neutrons with potential application in biology. Heinrich Stuhrmann has made pioneering contributions in this area (*127*).

4.7 The impacts of the SRS and ESRF in macromolecular crystallography

An objective measure of impact in the macromolecular crystallography field is *via* the SR facility and specific beamline statistics recorded at the website http://biosync.rcsb.org/. Table 4.2 summarizes the SRS beamlines' performance in contributing to these structure depositions. Also included is BM14, detailed in this review, the first third-generation SR source MX beamline. The data presented here were compiled in October 2011 (see http://biosync.rcsb.org/) and are likely to be reasonably complete, since the SRS closed operations in August 2008. The SRS has delivered 1471 structures (2.9%) of the total of 51077 macromolecular crystal structures determined using radiation from synchrotrons around the world, as of October 2011; in turn the 51077 structures are 74.2% of all the X-ray crystallography depositions. The ESRF third-generation source, in comparison, integrated over about half as many years, but about three times more beamlines, has delivered 7745 (15.2%) of the world's synchrotron depositions of structures, i.e. at a rate therefore per beamline of about three times greater than the second-generation SRS, but also with generally much more complex (larger molecular weight) structures than at SRS.

In more general terms, of impact assessments, two recent review articles on macromolecular crystallography at SR sources have appeared, one surveying the current status and future developments (*128*) and the other describing a personal view of the impact of SR on macromolecular crystallography (*129*). My own overviews can be found in (*130*) and (*131*), presented at a researcher level and at a teaching level respectively, both recently updated in second edition volumes.

In this account of the evolution of SR and crystallography, an appropriate perspective is that of one of the authors of the 1971 *Nature* muscle diffraction paper (*19*) from which I quote (*132*):

> The most important application (of synchrotron radiation) for biology later proved to be protein crystallography. Early tests of protein diffraction on the DESY source (Harmsen, Leberman & Schulz, 1976) (*133*) showed improvements compared with conventional sources but the gains were limited. The flux was about ten times better than with a conventional source. At this stage one had failed to appreciate that the parallel collimation of the beam was giving an unusually good signal-to-noise ratio. This was the property of synchrotron radiation which ultimately made it the source of choice for all kinds of protein crystal data collection. At about the same time studies on the Stanford storage ring SPEAR (Phillips, Wlodawer, Shevitz & Hodgson, 1976) (*18*) showed gains for crystal diffraction even with a non-focusing monochromator which indicated that (tailored) storage ring sources were going to be of considerable importance in protein crystallography. These authors made use of the ability to 'tune' the wavelength across an absorption edge to demonstrate the potentialities of synchrotron radiation in exploiting the effects of anomalous dispersion.... Protein crystallographers, whose numbers of course vastly exceeded the muscle community, began to appreciate the high speed and convenience of data collection at the storage ring sources, and soon began to realize that their data were better too, since radiation-induced damage had less time to fully manifest itself. Also, it became possible to work with

Table 4.2 Structures in the PDB for which data were collected at the SRS. The ESRF BM14, detailed in this review article, is also listed. BM14 was the first third-generation source MX beamline and is still active today

Synchrotron beamline	Total number of structures in the PDB
SRS7.2	135
SRS9.6	538
SRS9.5	146
SRS14.1	235
SRS14.2	223
SRS10.1	181
SRS_Unknown	19
Total	1471
ESRF_BM14	802

much smaller crystals, especially as the advantages of rapidly freezing them (before data collection) became apparent. It was ironic that the muscle experiments, which to a considerable extent had driven the technology needed for the use of synchrotron radiation, became almost a victim of their own success, for it became increasingly difficult, though not impossible, to obtain beam time because of the pressure from other users!

4.8 Other relevant topics

I need to add that I obviously had to be selective with the contents of my Lecture, and also to a degree with this publication. This has meant that some favourites were missing in the Lecture through pressure of time. Thus, I did not have time to highlight other favourite user challenges which featured in the evolutionary changes of user expectations. One such on SRS 7.2 was the 80% solvent content purine nucleoside phosphorylase project (*134*) which helped to contribute to the launch of BioCryst Pharmaceuticals (http://www.biocryst.com/) and brought SRS a mention in *Scientific American* (*135*). Methods development included a collaboration between Daresbury Laboratory and Rutherford Laboratory for a Multi-Wire Proportional Chamber (MWPC), built at Rutherford (*136*), for SAXS and Small Angle Diffraction; this may even have been one of the first involvements that Rutherford Laboratory ever had with SR. The 1 mm anode wire pitch for the 20 cm active area diameter was a cutting-edge aspect of the design for this MWPC. This whole arm of the evolutionary tree of instrument development at SR sources of MWPCs for X-ray diffraction and crystallography is a significant one and has been reviewed by Prof. Rob Lewis (*137*) and Prof. Roger Fourme (*138*).

I also had no time to discuss the methods and instrumentation lines of research involving the potential and R&D work done so far on harnessing high photon energies in crystallography (ultrashort wavelengths in the range 0.3 Å) (*139*) and with its potential for charge density studies. This latter field is a growing discipline in structural chemistry but has not yet taken off really with SR, presumably because of its exacting demands on SR source stability. This domain and the use of high photon energies have also been discussed by Coppens (*140*) and by Hart (*141*). A beamline in the final phase of SRS was beamline 10, for structural genomics (a project led by Prof. Samar Hasnain) (*142*). Also there is protein powder diffraction at ESRF ID31 for which we have started to make a contribution (*143*), as one frontier in the protein crystallography field as it moves to individual, ever smaller, crystals even below 1 μm (*144*), as well as polycrystallography (several single crystals in the X-ray beam at once (*145*)) and bulk protein powder diffraction (reviewed in (*146*)).

Nor have I covered adequately the applications of resonant scattering in chemical crystallography, another favourite theme in our laboratory in Manchester, led by Dr Madeleine Helliwell, showing also the synergies between biological and chemical crystallography methods developments (*147, 148*).

Neither did I have time to highlight the important work of the Daresbury Analytical Research and Technical Services (DARTS) (see ref. (*149*) for a recent summary) nor the economic and social impact of the Daresbury SRS as a whole (*150*). These aspects are of course important, not least with the increasingly wide, albeit still controversial, recognition now of the importance of the 'impact' of the science that we undertake. From the outset of the SRS, I recall explaining the potential impact of these SRS instrumentation developments, for example at the conference in Nordwijk in the Netherlands on Medical Applications of Synchrotron Radiation (*151*).

The visits to various excellent overseas SR facilities by myself or by members of my group or close colleagues have also not been mapped out due to lack of time in the Lecture, but which involved methods and facility development, besides SRS or ESRF, such as for reciprocal-space mapping at the NSLS (*152*), resonant scattering crystallography at NSLS and ELETTRA (*153, 154*). Visits as a user to APS IMCA CAT, APS SBC CAT, SOLEIL, DORIS and PETRA III have also not been described.

At IUCr Madrid (2011), I reported on a current study (N. Chayen, L. Govada, J. R. Helliwell and S. Tanley to be published) involving experiments using a micro-beam (about 10 μm diameter) scanned across an α-crustacyanin crystal at the DLS macromolecular crystallography microfocus facility (*144*), which we have undertaken to search for the best ordered portion.

In terms of patents, I could not find in my patent searches any patent attributed to Kathleen Lonsdale. My efforts at securing patents have led to one (the 3D 'toast-rack' Laue diffraction film or image plates arrangement (*155*)) and contributed to one other, but neither have made it big, and so patents remain a future priority objective for my research.

The fascination I have with the foundations of crystallography includes symmetry and all the manifest deviations from perfect symmetry that can occur in Nature, and that we study or find workarounds for crystal structure analysis (*156, 157*). There are also the crystals themselves and crystal growth phenomena and their monitoring and evaluation.

After my Lecture there was one question, from Dr Sax Mason of the Institut Laue Langevin in Grenoble (*158*): 'With all this improved SR source brightness that we are still witnessing with improved sources is it leading to a higher rate of deposition in the PDB (Protein Data Bank)?'. I replied: 'Yes and no; there are increases but a user can now be more selective in choosing whichever sample to proceed to data collection with, helped also by widespread availability of robotics for automatic sample changing at beamlines'. In addition, I could have said that projects are getting more ambitious, such as time-resolved experiments and larger complexes (e.g. viruses) and the ribosome projects. Indeed a careful, systematic, analysis of SR versus laboratory source X-ray data determined macromolecular crystal structures deposited in the PDB has been made, emphasizing, among a variety of conclusions, the larger molecular weight on average of the SR-based studies (*159*). However, there are also, increasingly, complications with X-radiation damage, even with cryo-cooled crystals (*160*), and several volumes of *Journal of Synchrotron Radiation* have been devoted to the publications arising from several Workshops led by Prof. Elspeth Garman, Dr Colin Nave and Dr Sean McSweeney (*161*).

4.9 Research directions for the future

It is invidious to pick out particular directions, but I select some personal favourites in this simple list below.

- New SR sources: PETRA III, recently on-line, is currently the most brilliant, i.e. the 'ultimate', storage ring.
- New SR sources: NSLSII, MAX IV and ALBA all will offer their users marvellous levels of specifications as SR sources.
- Our own national UK DLS is a 'cutting edge' source and suite of instruments and also with a vibrant research and development programme. Figure 4.15 shows the beamline layout for DLS I19 the chemical crystallography beamline, by way of illustration of a modern state-of-the-art beamline for crystallography.
- ESRF has a new upgrade programme for nanofocus beams; UK has a 10% share in ESRF. APS and SPRing-8 also are embarking on major upgrade programmes.
- Structural biology will show a continued expansion into studying large complexes combined with small angle X-ray scattering (SAXS) and electron microscopy (EM); our own example in this area is with the multi-macromolecular complex crustacyanin (*162*).
- Structural chemistry will I believe show a continued expansion into time-resolved crystallography (*163, 164*).
- The X-ray lasers user programme will strive towards reaching the goal of single-molecule diffraction.
- New and upgraded neutron sources in Europe, United States and Japan will benefit crystallography, reaching full hydrogenation details completed for mechanistic and functional studies, as well as molecular recognition studies.

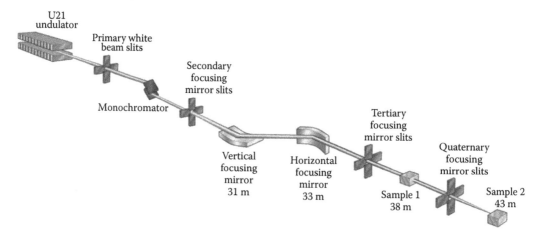

Figure 4.15 **(See colour insert.)** The layout of the DLS I19 chemical crystallography beamline. With permission of DLS and Professor Dave Allan, Scientist in charge.

The initiatives in SR biological diffraction at DESY (*19*) and in SR protein crystallography at Stanford SSRL (*18*) allowed scientists like myself to join in the excitement and help bring about the profound changes in technical and scientific capabilities that SR has brought to the field of macromolecular crystallography, and indeed all of crystallography.

Acknowledgements

I am grateful to Prof. Trevor Greenhough and the BCA Keele 2011 Conference Programme Committee for the invitation to present the 2011 Lonsdale Lecture. I am grateful to all my coauthors and collaborators and PhD students, for all our joint work together during the decades. To the University of Manchester (since 1989) and Daresbury Laboratory (since 1976), also to the ESRF and the Institut Laue Langevin for stimulating environments, as well as my stays at the Universities of York, Oxford, and Keele Universities; and finally to the various funding agencies, a heartfelt 'thank you!'. I am grateful to Kate Crennell for help with the historical record of previous Lonsdale Lecturers. After my Lecture, I circulated my PowerPoint file to various colleagues that I have pictured such as Howard Einspahr, Britt Hedman, Keith Hodgson and Keith Moffat and their positive comments I was pleased to receive. The preparation of my Lecture was greatly assisted by my wife and colleague, Dr Madeleine Helliwell, to whom I am very grateful. The 'Durward Cruickshank Letters Archive' donated by Durward to the University of Manchester, and now properly listed and looked after at the John Rylands University of Manchester Library (JRULM) and which can be accessed *via* the JRULM Archivist, was largely the work of the science librarian John Blunden-Ellis of the Joule Library, University of Manchester, to whom thanks are due. I am grateful to Professor Michael Hart for extensive discussions concerning the early 1970s development of an SR policy for the UK as well as for detailed subsequent comparisons and insights regarding the NSLS and SRS projects, and also to Prof. Sir Ron Mason for his permission to reproduce his 1975 letter, which proved so influential in my career development. I am grateful to Dr Gwyn Williams of Jefferson Laboratory, United States, for kindly providing Figure 4.5. I am also grateful to the various authors and journals who have responded promptly to my copyright permissions requests, and I am grateful to the following colleagues who provided comments (in whole or in part) on this review article draft: Prof. André Authier, Dr Matthew Blakeley, Prof. Naomi Chayen, Mr Mike Dacombe, Prof. Trevor Forsyth, Prof. Samar Hasnain, Dr Sean McSweeney, Dr Judith Milledge, Dr Colin Nave, Dr Pierre Rizkallah, Prof. Francesco Sette, Mr Peter Strickland and Dr Andrew Thompson. I am grateful to the two anonymous referees and the Editor, Prof. Moreton Moore, who each provided constructive criticisms. Finally, I would like to acknowledge the patience of SERC, which became

Council for the Central Laboratory of the Research Councils (CCLRC) and finally became STFC, in accommodating me in various employment modes between 1979 and 2010 (full-time or part-time as a Joint Appointment, and finally as an Honorary Visiting Scientist at Daresbury Laboratory).

Notes on the contributor

This photo of the author in the SRS 7.2 experimental hutch was taken by Stuart Eyres, Daresbury Laboratory, in 2002 and is reproduced with the permission of STFC Daresbury Laboratory. J.R. Helliwell is, since 1989, Professor of Structural Chemistry of the University of Manchester. John Helliwell worked at Daresbury Laboratory's SRS from 1979 to 1993, whilst also a Joint Appointee with the Universities of Keele, York and Manchester, and full time as a scientific civil servant (1983–1985), and earlier on the NINA synchrotron (1976). He was Director of SR Science full-time at CCLRC based at Daresbury Laboratory in 2002. In 2002, he was awarded the Banerjee Centennial Silver Medal of the Indian Association for the Cultivation of Science in Calcutta, India; his Banerjee Centennial Lecture was published (*165*). He was part-time Science Adviser to the Daresbury Analytical and Technical Research Services 'DARTS', then STFC CLIK, from 2003 to 2009. He has served on the ESRF Science Advisory Committee as Vice-Chairman and then Chairman, on the ESRF MAC, and on the Council of the ESRF; as Chairman of the UK SRS Panel for Protein Crystallography; as Chairman of the Cornell University Macromolecular Crystallography at CHESS (MACCHESS) Advisory Committee; as a Member of the Elettra Sincrotrone Trieste Review Committee; as a Reviewer of the EMBL Outstation in Hamburg and on the Science Advisory Committee of the Advanced Photon Source. He was the founding Chairman of the International Union of Crystallography's Commission on Synchrotron Radiation. He was one of the founding Editors of the *Journal of Synchrotron Radiation,* an Editor of the OUP Book Series on SR and Editor-in-Chief of *Acta Crystallographica* from 1996 to 2005. He served as President of the European Crystallographic Association from 2006 to 2009. He is currently serving on the Advisory Committee for the Australian Research Council Centre for Coherent X-ray Science and is Chairman of the Spanish Synchrotron Source 'ALBA' Science Advisory Committee.

Summary of abbreviations used

APS – Advanced Photon Source at Argonne National Laboratory
CCLRC – Council for the Central Laboratory of the Research Councils
CHESS – Cornell High Energy Synchrotron Source
DAFS – Diffraction Anomalous Fine Structure
DESY – Deutsches Elektronen-Synchrotron
DLS – Diamond Light Source

EMBL – European Molecular Biology Laboratory
ESRF – European Synchrotron Radiation Facility
ESRP – European Synchrotron Radiation Project
HBL – High Brightness Lattice
LURE – Laboratoire pour l'utilization du rayonnement électromagnétique
MacCHESS – Macromolecular Crystallography at CHESS
MAD – Multiple-Wavelength Anomalous Dispersion
NSLS – National Synchrotron Light Source at Brookhaven
SERC – Science and Engineering Research Council
SLAC – Stanford Linear Accelerator Centre
SPRing-8 – Super Photon Ring 8 GeV
SRS – Synchrotron Radiation Source
SR – Synchrotron Radiation
SSRL – Stanford Synchrotron Radiation Laboratory
STFC – Science and Technological Facilities Council

References

[1]   Hodgkin, D. Kathleen Lonsdale 28 January 1903–1 April 1971. *Biogr. Mems Fell. R. Soc.* **1975,** *21,* 447–484.

[2]   (a) Announcement of the Kathleen Lonsdale Lectures. *Acta Cryst.* (1987) A43, 589 Notes and News; (b) Watkin, D.J. Chemical Crystallography – Science, Technology or a Black Art. *Crystallogr. Rev.* 2010, 16, 197–230.

[3]   Lonsdale, K. *Crystals and X-rays*; G. Bell & Sons: London, 1948.

[4]   Buerger, M.J. Book Review of Crystals and X-rays. *Acta Cryst.* **1950,** *3,* 79.

[5]   Lonsdale, K. An X-ray Analysis of the Structure of Hexachlorobenzene, using the Fourier Method. *Proc. R. Soc. Lond. A* **1931,** *133,* 536–552.

[6]   Lonsdale, K. The Structure of the Benzene Ring in $C_6(CH_3)_6$. *Proc. R. Soc. Lond.* **1929,** *A123,* 494–515.

[7]   Anonymous Obituary of Kathleen Lonsdale *The Times* newspaper, 1971.

[8]   Extract of Letter from Kathleen Lonsdale to the Ministry of Labour and National Services, 1942.

[9]   Lonsdale, K. *Is Peace Possible?* Penguin: Harmondsworth, Middlesex, 1957; p. 127.

[10]   Lonsdale, K.; Milledge, J. Analysis of Thermal Vibrations in Crystals: A Warning. *Acta Cryst.* **1961,** *14,* 59.

[11]   Cruickshank, D.W.J. Tilting at Windmills 2005 *IUCr Newsletter*, http://www.iucr.org/news/newsletter/volume-13/number-2/tilting-at-windmills

[12]   Bragg, W.L. Kathleen Lonsdale. *Acta Cryst.* **1972,** *A28,* 226.

[13]   Extract from an email Dr Judith Milledge to the author July 2011.

[14]   Milledge, J. Kathleen Yardley Lonsdale (1903–1971). In *Out of the shadows*; Byers, N., Williams, G., Eds.; C.U.P.: Cambridge, 2006; pp. 191–201.

[15]   Milledge, J. Obituary of Kathleen Lonsdale 1903–1971. *Acta Cryst.* **1975,** *A31,* 705–708.

[16]   Bordas, J.; Munro, I.H.; Glazer, A.M. Small-angle Scattering Experiments on Biological Materials using Synchrotron Radiation. *Nature* **1976,** *262,* 541.

[17]   Letter from Sir Ron Mason to Sir Sam Edwards (July 2nd 1975).

[18]   Phillips, J.C.; Wlodawer, A.; Yevitz, M.M.; Hodgson, K.O. Applications of Synchrotron Radiation to Protein Crystallography: Preliminary Results. *PNAS* **1976,** *73,* 128–132.

[19]   Rosenbaum, G.; Holmes, K.C.; Witz, J. Use of SR in biological diffraction *Nature (London),* **1971,** *230,* 434–437.

[20]   Helliwell, J.R. In Optimisation of Anomalous Scattering and Structural Studies of Proteins using Synchrotron Radiation, Proceedings of the Daresbury Study Weekend, Jan 26–28, 1979; DL/SCI/R13; 1979 pp 1–6.

[21]   Phillips, D.C. Quoted in The Scientific Case for Research with Synchrotron Radiation SRC Daresbury Lab DL/SRF/R3, 1975.

[22] Hayes, W. (Oxford U.), Beaumont, J.H. (Oxford U.), GA Brooker (Oxford U.), MJ Cooper (Warwick U.), M Hart (Bristol U.), HE Huxley (Cambridge U.), Professor D.C. Phillips (University of Oxford) Dr. K.R. Lea (Daresbury Laboratory). *Report of the Crystal Optics Panel for Instrumentation on the SRS DL Technical Reports,* DL-SRF-TM08, 1976.

[23] Beaumont, J.H.; Hart, M. Multiple Bragg Reflection Monochromators for Synchrotron Radiation. *J. Phys. E Sci. Instrum.* **1974,** *7,* 823–829.

[24] Hart, M. Synchrotron Radiation – Its Application to High-Speed, High-Resolution X-ray Diffraction Topography. *J. Appl. Cryst.* **1975,** *8,* 436–444.

[25] Beaumont, J.H.; Grime, G.W.; Hart, M. A High Resolution, High Intensity Small Angle Scattering Camera for Synchrotron x-radiation. *J. Phys. E Sci. Instrum.* **1976,** *9,* 680–683.

[26] Munro, I.H. Synchrotron Radiation Research in the UK. *J. Synch. Rad.* **1997,** *4,* 344–358.

[27] Le Lay, G.; Lindley, P.; Margaritondo, G. Future Applications of Science with Synchrotron Radiation and Free-electron Lasers in Europe. *J. Synch. Rad.* **2001,** *8,* 1167–1168.

[28] Helliwell, J.R. Global Instrumentation Survey of the IUCr Commission on Synchrotron Radiation I Macromolecular Crystallography. *Synch. Rad. News* **1992,** *5,* No 2, 22–27.

[29] Cernik, R.J.; Helliwell, J.R.; Poole, M.; Turner, T. Meeting Report SRS Highlights Presented at the British Association Festival of Science. *Synch. Rad. News* **2009,** *22* (4), 10–12.

[30] Hasnain, S.S.; Helliwell, J.R. (Report on) Journal of Synchrotron Radiation. *Acta Cryst.* **1993,** *A49* (Suppl.), c411–c412.

[31] Hasnain, S.S.; Helliwell, J.R.; Kamitsubo, H. Overview on Synchrotron Radiation and the Need for the Journal of Synchrotron Radiation. *J. Synch. Rad.* **1994,** *1,* 1–4.

[32] Authier, A. 60 years of IUCr Journals. *Acta Cryst.* **2009,** *A65,* 167–182.

[33] Helliwell., J.R. Triennial Report (1990–1993) to the IUCr Executive Committee as Chairman of the Commission on Synchrotron Radiation. *Acta Cryst.* **1995,** *A51,* 622.

[34] Hasnain, S.S.; Helliwell, J.R.; Kamitsubo, H. JSR: Required Reading. *J. Sync, Rad.* **1996,** *3,* 247.

[35] Hasnain, S.S.; Helliwell, J.R.; Kamitsubo, H. A Perspective: *JSR* so far. *J. Synch. Rad.* **1995,** *2,* 275.

[36] Hasnain, S.S.; Helliwell, J.R.; Kamitsubo, H. Synchrotron Radiation Sources in the News and *JSR* goes Online. *J. Synch. Rad.* **1999,** *6,* 1069–1070.

[37] *Nature* 1996 383, 42.

[38] Hasnain, S.S.; Kamitsubo, H.; Mills, D.M. New Synchrotron Radiation Sources and the Next-generation Light Sources. *J. Synch. Rad.* **2001,** *8,* 1171.

[39] Schwinger, J. On the Classical Radiation of Accelerated Electrons. *Phys. Rev.* **1949,** *75,* 1912–1925.

[40] Helliwell, J.R. Synchrotron Radiation and Crystallography: The First Fifty Years. *Acta Cryst. A* **1998,** *A54,* 738–749.

[41] The ESRF Foundation Phase Report (1987) Grenoble, France.

[42] Stefan, P.; Krinsky, S.; Rakowsky, G.; Solomon, L. Small-gap Undulator Experiment on the NSLS X-ray Ring, Proceedings of the 1995 Particle Accelerator Conference, Dallas, TX, 1995. IEEE 95CH35843.

[43] Garner, C.D.; Helliwell, J.R. Uses of SR in Biochemical Research. *Chem. Brit.* **1986,** *22,* 835–840 [and the front cover].

[44] Helliwell, J.R.; Worgan, J.S. Optimal Geometry of a Curved Crystal Monochromator for Protein Crystallography with Synchrotron Radiation, *SERC Daresbury Laboratory Preprint,* 1979 DL/SCI/P203E.

[45] Lemonnier, M.; Fourme, R.; Rousseaux, F.; Kahn, R. X-ray Curved-crystal Monochromator System at the Storage Ring DCI. *Nucl. Instrum. Methods* **1978,** *152* (1), 173–177.

[46] Greenhough, T.J.; Helliwell, J.R. Oscillation Camera Data Processing: Reflecting Range and Prediction of Partiality: I. Conventional Sources. *J. Appl. Cryst.* **1982,** *15,* 338–351.

[47] Greenhough, T.J.; Helliwell, J.R. Oscillation Camera Data Processing: Reflecting Range and Prediction of Partiality: II. Synchrotron Sources. *J. Appl. Cryst.* **1982,** *15,* 493–508.

[48] Helliwell, J.R.; Greenhough, T.J.; Carr, P.; Rule, S.A.; Moore, P.R.; Thompson, A.W.; Worgan, J.S. Central Data Collection Facility for Protein Crystallography, Small Angle Diffraction and Scattering at the Daresbury SRS. *J. Phys. E.* **1982,** *15,* 1363–1372.

[49] Arndt, U.W.; Greenhough, T.J.; Helliwell, J.R.; Howard, J.A.K.; Rule, S.A.; Thompson, A.W. Optimised Anomalous Dispersion Crystallography: A Synchrotron X-ray Polychromatic Simultaneous Profile Method. *Nature* **1982,** *298,* 835–838.

[50] Helliwell, J.R. Synchrotron X-radiation Protein Crystallography: Instrumentation, Methods and Applications. *Rep. Progr. Phys.* **1984,** *47,* 1403–1497.

[51] Greenhough, T.J.; Helliwell, J.R.; Rule, S.A. Oscillation Camera Data Processing: III. Spot Shape, Size and Energy Profile. *J. Appl. Cryst* **1983,** *16,* 242–250.

[52] Stragier, H.; Cross, J.O.; Rehr, J.J.; Sorensen, L.B.; Bouldin, C.E.; Woicik., J.C. Diffraction Anomalous Fine Structure: A New X-ray Structural Technique. *Physical Review Letters* **1992,** *69* (21), 3064–3067.

[53] Vacinova, J.; Hodeau, J.L.; Wolfers, P.; Lauriat, J.P.; ElKaim, E. Use of Anomalous Diffraction, DAFS and DANES Techniques for Site-selective Spectroscopy of Complex Oxides. *J. Synch. Rad.* **1995,** *2* (5), 236–244.

[54] Rossmann, M.G.; Erickson, J.W. Oscillation Photography of Radiation-sensitive Crystals using a Synchrotron Source. *J. Appl. Cryst.* **1983,** *16,* 629–636.

[55] Einspahr, H.; Suguna, K.; Suddath, F.L.; Ellis, G.; Helliwell, J.R.; Papiz., M.Z. The Location of the Mn:Ca Ion Cofactors in Pea Lectin Crystals by Use of Anomalous Dispersion and Tunable Synchrotron X-radiation. *Acta Cryst.* **1985,** *B41,* 336–341.

[56] Cianci, M.; Rizkallah, P.J.; Olczak, A.; Raftery, J.; Chayen, N.E.; Zagalsky, P.F.; Helliwell, J.R. Structure of Apocrustacyanin A1 using Softer X-rays. *Acta Crystallogr. Sect. D-Biol. Crystallogr.* **2001,** *D57,* 1219–1229.

[57] Helliwell, J.R. Overview and New Developments in Softer X-ray ($2\text{Å} < \lambda < 5\text{Å}$) Protein Crystallography. *J. Synch. Rad.* **2004,** *11,* 1–3.

[58] Moffat, K.; Szebenyi, D.; Bilderback, D. X-ray Laue Diffraction from Protein Crystals. *Science.* **1984,** *223* (4643), 1423–5.

[59] Wyckoff, Harold W.; Doscher, Marilynn; Tsernoglou, Demetrius; Inagami, Tadashi; Johnson, Louise N.; Hardman, Karl D.; Allewell, Norma M.; Kelly, David M.; Richards, Frederic M. Design of a Diffractometer and Flow Cell System for X-ray Analysis of Crystalline Proteins with Applications to the Crystal Chemistry of Ribonuclease-S. *J. Mol. Biol.* **1967,** *27,* 563–578.

[60] Helliwell, J.R. The Use of Electronic Area Detectors for Synchrotron X-radiation Protein Crystallography with Particular Reference to the Daresbury SRS. *Nucl. Instrum. Methods.* **1982,** *201,* 153–174.

[61] Arndt, U.W.; Gilmore, D.J. X-ray Television Area Detectors for Macromolecular Structural Studies with Synchrotron Radiation Sources. *J. Appl. Cryst.* **1979,** *12,* 1–9.

[62] Helliwell, J.R.; Papiz, M.Z.; Glover, I.D.; Habash, J.; Thompson, A.W.; Moore, P.R.; Harris, N.; Croft, D.; Pantos, E. The Wiggler Protein Crystallography Work-station at the Daresbury SRS; Progress and Results. *Nucl. Instrum. Methods.* **1986,** *A246,* 617–623.

[63] Baker, P.J.; Farrants, G.W.; Stillman, T.J.; Britton, K.L.; Helliwell, J.R.; Rice, D.W. Isomorphous Replacement with Optimised Anomalous Scattering Applied to Protein Crystallography. *Acta Cryst.* **1990,** *A46,* 721–725.

[64] Hedman, B.; Hodgson, K.O.; Helliwell, J.R.; Liddington, R.; Papiz, M.Z. Protein Micro-crystal Diffraction and the Effects of Radiation Damage with Ultra High Flux Synchrotron Radiation. *PNAS. USA* **1985,** *82,* 7604–7607.

[65] Helliwell, J.R.; Habash, J.; Cruickshank, D.W.J.; Harding, M.M.; Greenhough, T.J.; Campbell, J.W.; Clifton, I.J.; Elder, M.; Machin, P.A.; Papiz, M.Z.; Zurek, S. The Recording and Analysis of Laue Diffraction Photographs. *J. Appl. Cryst.* **1989,** *22,* 483–497.

[66] Pickersgill, R.W. One-dimensional Disorder in Spinach Ribulose Bisphosphate Carboxylase Crystals. *Acta Cryst.* **1987,** *A43,* 502–506.

[67] Hajdu, J.; Acharya, K.R.; Stuart, D.I.; McLaughlin, P.J.; Barford, D.; Oikonomakos, N.G.; Klein, H.W.; Johnson., L.N. Catalysis in the Crystal: Synchrotron Radiation Studies with Glycogen Phosphorylase b. *EMBO J.* **1987,** *6,* 539–546.

[68] Hajdu, J.; Machin, P.A.; Campbell, J.W.; Greenhough, T.J.; Clifton, I.J.; Zurek, S.; Gover, S.; Johnson, L.N.; Elder, M. Millisecond X-ray Diffraction and the First Electron Density Map from Laue Photographs of a Protein Crystal. *Nature* **1987,** *329,* 178–181.

[69] Acharya, R.; Fry, E.; Stuart, D.; Fox, G.; Rowlands, D.; Brown, F. The Structure of Foot-and-mouth Disease Virus at 2.9 Å. *Nature (London)* **1989,** *337,* 709–716.

[70] Liddington, R.C.; Yan, Y.; Moulai, J.; Sahli, R.; Benjamin, T.L.; Harrison, S.C. Structure of Simian Virus 40 at 3.8Å Resolution. *Nature (London)* **1991,** *354,* 278–284.

[71] Yonath, A.; Harms, J.; Hansen, H.A.S.; Bashan, A.; Schlünzen, F.; Levin, I.; Koelln, I.; Tocilj, A.; Agmon, I.; Peretz, M.; Bartels, H.; Bennett, W.S.; Krumholz, S.; Janell, D.; Weinstein, S.; Auerbach, T.; Avila, H.; Piolleti, M.; Morlang, S.; Franceschi, F. Crystallographic Studies on the Ribosome, a Large Macromolecular Assembly Exhibiting Severe Nonisomorphism, Extreme Beam Sensitivity and No Internal Symmetry. *Acta Cryst.* **1998,** *A54,* 945–955.

[72] Ban, N.; Nissen, P.; Hansen, J.; Capel, M.; Moore, P.B.; Steitz, T.A. Placement of Protein and RNA Structures into a 5 Å-Resolution Map of the 50S Ribosomal subunit. *Nature* **1999,** *400,* 841–847.

[73] Wonacott, A.J.; Skarzinski, T. Glaxo (commercial work; unpublished).

[74] Andrews, S.J.; Papiz, M.Z.; McMeeking, R.; Blake, A.J.; Lowe, B.M.; Franklin, K.R.; Helliwell, J.R.; Harding, M.M. Piperazine Silicate (EU-19) – The Structure of a Very Small Crystal Determined with Synchrotron Radiation. *Acta Cryst.* **1988,** *B44,* 73–77.

[75] Harding, M.M. Synchrotron Radiation – New Opportunities for Chemical Crystallography. *Acta Cryst.* **1995,** *B51,* 432–446.

[76] Clegg, W. Synchrotron Chemical Crystallography. *J. Chem. Soc., Dalton Trans.,* **2000,** 3223–3232.

[77] Duke, E.M.H.; Kehoe, R.C.; Rizkallah, P.J.; Clarke, J.A.; Nave, C. Beamline 14: A New Multipole Wiggler Beamline for Protein Crystallography on the SRS. *J. Synch. Rad.* **1998,** *5,* 497–499.

[78] Thompson, D.J. (1985) In The Origins, and the Construction, Commissioning and Operation of the SRS p50–74 in Construction and Commissioning of Dedicated Synchrotron Radiation Facilities, Proceedings of a Workshop held Oct 16–18, 1985 at Brookhaven National Laboratory, Roger W. Klaffky Ed., USA Report number BNL 51959.

[79] Thompson, D.J.; Suller, V.P. Conversion of the SRS to a Higher Brilliance Lattice. *Rev. Sci. Instrum.* **1989,** *60,* 1377–1381.

[80] Hope, H. Cryocrystallography of Biological Macromolecules: A Generally Applicable Method. *Acta Crystallogr* **1988,** *B44,* 22–26.

[81] Haas, D.J.; Rossmann, M.G. Crystallographic Studies On Lactate Dehydrogenase at −75°C. *Acta Cryst* **1970,** *B26,* 998–1004.

[82] Mitchell, E.P.; Garman, E.F. Flash Freezing of Protein Crystals: Investigation of Mosaic Spread and Diffraction Limit with Variation of Cryoprotectant Concentration. *J. Appl. Cryst.* **1994,** *27,* 1070–1074.

[83] Brammer, R.C.; Helliwell, J.R.; Lamb, W.; Liljas, A.; Moore, P.R.; Thompson, A.W.; Rathbone, K. A New Protein Crystallography Station on the SRS Wiggler Beamline for Very Rapid Laue and Rapidly Tunable Monochromatic Experiments: I. Design Principles, Ray Tracing and Heat Calculations. Nucl. Instrum. *Methods* **1988,** *A271,* 678–687.

[84] Peterson, M.R.; Harrop, S.J.; McSweeney, S.M.; Leonard, G.A.; Thompson, A.W.; Hunter, W.N.; Helliwell, J.R. MAD Phasing Strategies Explored with a Brominated Oligonucleotide Crystal at 1.65Å Resolution. *J. Synch. Rad.* **1996,** *3,* 24–34.

[85] Maier-Leibnitz, H. (Chair). *European Science Foundation Working Group on Synchrotron Radiation Synchrotron Radiation a Perspective View for Europe 1977;* ESF Report No. 3.000/77, 87 pp.

[86] Farge, Y., Ed. *The European Synchrotron Radiation facility: The Feasibility Study.* The European Science Foundation, 1979, ISBN 2-903148-01-5, 66 pp.

[87] Farge, Y.; Duke, P.J., Eds. *The European Synchrotron Radiation facility: Supplement I.* The Scientific Case European Science Foundation, ISBN 2-903148-02-3, (1979) 173 pp.

[88] Thompson, D.J.; Poole, M.W. The European Synchrotron Radiation facility: Supplement II The Machine 1979 European Science Foundation ISBN 2-903148-03-1, 159 pages.

[89] Buras, B.; Marr, G.V. The European Synchrotron Radiation facility: Supplement III Instrumentation European Science Foundation ISBN 2-903148-04-X, 180 pages.

[90] (a) Helliwell J.R.; Farge, Y.; Duke, P.J., Eds. The European Synchrotron Radiation facility: Supplement I The Scientific Case European Science Foundation ISBN 2-903148-02-3, page 116. (b) Wilson, K.S.; Stura, E.A.; Wild, D.L.; Todd, R.J.; Stuart, D.I.; Babu, Y.S.; Jenkins, J.A.; Standing, T.S.; Johnson, L.N.; Fourme, R.; Kahn, R.; Gadet, A.; Bartels, K.S.; Bartunik, H.D. Macromolecular Crystallography with Synchrotron Radiation. II. Results. *J. Appl. Cryst.* (1983). 16, 28–41.

[91] Buras, B. 1985 The European Synchrotron Radiation Facility pages 195–204 in Construction and Commissioning of Dedicated Synchrotron Radiation Facilities, Proceedings of a Workshop held October 16–18, 1985 at Brookhaven National Laboratory, USA Report number BNL 51959, Editor Roger W. Klaffky.

[92]   (a) Helliwell, J.R. Protein Crystal Perfection and the Nature of Radiation Damage, (Paper presented at the 2nd International Conference on Protein Crystal Growth, Bischenberg, Alsace, July 1987), *J. Crystal Growth* **1988**, *90*, 259–272; (b) Hedman, B.; Hodgson, K.O.; Helliwell, J.R.; Liddington R. unpublished.

[93]   Helliwell, J.R.; Fourme, R. The ESRF as a Facility for Protein Crystallography: A Report and Design Study, ESRP Report IRI-4/83, 1983, pp 1–36.

[94]   Kam, Z. Chapter 5 'Measurement of spatial correlations of fluctuations in synchrotron X-ray scattering from solutions' p. 103 in Stuhrmann, H.B. Ed. Uses of Synchrotron Radiation in Biology; Academic Press: London, 1982, pp xii + 348.

[95]   Kuzay, T.M.; Kazmierczak, M.; Hsieh, B.J. X-ray Beam/Biomaterial Thermal Interactions in Third-generation Synchrotron Sources. *Acta Cryst.* **2001**, *D57*, 69–81.

[96]   Snell, E.H.; Bellamy, H.D.; Rosenbaum, G.; van der Woerd, M.J. Non-invasive Measurement of X-ray Beam Heating on a Surrogate Crystal Sample. *J. Synch. Rad.* **2007**, *14*, 109–115.

[97]   Miller, A. ESRF Newsletter, Grenoble, France.

[98]   Riekel, C.; Burghammer, M.; Schertler, G. Protein Crystallography Microdiffraction. *Curr. Opin. Struct. Biol.* **2005**, *15*, 556–562.

[99]   Perrakis, A.; Cipriani, F.; Castagna, J.-.C.; Claustre, L.; Burghammer, M.; Riekel, C.; Cusack, S. Protein Microcrystals and the Design of a Microdiffractometer: Current Experience and Plans at EMBL and ESRF/ID13. *Acta Cryst.* **1999**, *D55*, 1765–1770.

[100] Holton, J.M.; Frankel, K.A. The Minimum Crystal Size Needed for a Complete Diffraction Data Set. *Acta Cryst.* **2011**, *D66*, 393–408.

[101] Nave, C.; Hill., M.A. Will Reduced Radiation Damage Occur with Very Small Crystals? *J. Synch. Rad.* **2005**, *12*, 299–303.

[102] Hastings, J.B. NSLS EXAFS Beamline Design: Collimating Mirrors for X-ray Beamlines, Brookhaven National Laboratory Report BNL-30617.

[103] Helliwell, J.R. Macromolecular Crystallography with Synchrotron Radiation Cambridge University Press 1992. Published in paperback 2005. Cambridge.

[104] Hendrickson, W.; Horton, J.R.; LeMaster, D.M. Selenomethionyl Proteins Produced for Analysis by Multiwavelength Anomalous Diffraction (MAD): A Vehicle for Direct Determination of Three-dimensional Structure. *EMBO J* **1990**, *9*, 1665–1672.

[105] Haedener, A.; Matzinger, P.K.; Battersby, A.R.; McSweeney, S.; Thompson, A.W.; Hammersley, A.P.; Harrop, S.J.; Cassetta, A.; Deacon, A.; Hunter, W.N.; Nieh, Y.P.; Raftery, J.; Hunter, N.; Helliwell, J.R. Determination of the Structure of Seleno-methionine-labelled Hydroxymethylbilane Synthase in its Active Form by Multi-wavelength Anomalous Dispersion. *Acta Cryst* **1999**, *D55*, 631–643.

[106] Bragg, W.L. *Development of X-ray Analysis*; Chapter 11 Methods of Measurement; Phillips, D.C., Lipson, H., Eds.; Bell: London, 1975; p. 137.

[107] Wyckoff, R.W.G. The Crystal Structures of Some Carbonates of the Calcite Group. *Am. J. Sci.* **1920**, *50*, 317–360.

[108] Bragg, W.L. The Structure of Some Crystals as Indicated by their Diffraction of X-rays. *Proc. R. Soc. Lond. A* **1913**, *89*, 248–277.

[109] Campbell, J.; Habash, J.; Helliwell, J.R.; Moffat, K. Wavelength Normalisation of Laue Data, Information Quarterly No. 19 SERC, Daresbury Laboratory (1986).

[110] Cruickshank, D.W.J.; Helliwell, J.R.; Moffat, K. Multiplicity Distribution of Reflections in Laue Diffraction. *Acta Cryst.* **1987**, *A43*, 656–674.

[111] Cruickshank, D.W.J.; Helliwell, J.R.; Moffat, K. Angular Distribution of Reflections in Laue Diffraction. *Acta Cryst.* **1991**, *A47*, 352–373.

[112] Helliwell, J.R.; Harrop, S.; Habash, J.; Magorrian, B.G.; Allinson, N.M.; Gomez, D.; Helliwell, M.; Derewenda, Z.; Cruickshank, D.W.J. 'Instrumentation for Laue diffraction' (1989) *Rev. Sci. Instrum.* 60(7), 1531–1536 (Invited paper at the International Conference on Synchrotron Radiation Instrumentation held Tsukuba, September 1988).

[113] Srajer, V.; Teng, T.; Ursby, T.; Pradervand, C.; Ren, Z.; Adachi, S.; Schildkamp, W.; Bourgeois, D.; Wulff, M.; Moffat, K. Photolysis of the Carbon Monoxide Complex Of Myoglobin: Nanosecond Time-resolved Crystallography. *Science* **1996**, *274* (5293), 1726–9.

[114] Moy, J.P. A 200 mm Input Field, 5–80 keV Detector based on an X-ray Image Intensifier and CCD Camera. *Nucl. Instrum. Methods* **1994**, *A348*, 641–644.

[115] Helliwell, J.R.; Nieh, Y.P.; Raftery, J.; Cassetta, A.; Habash, J.; Carr, P.D.; Ursby, T.; Wulff, M.; Thompson, A.W.; Niemann, A.C.; Hádener, A. Time-resolved Structures of Hydroxymethylbilane Synthase (Lys59Gln mutant) as it is Loaded with Substrate in the Crystal Determined by Laue Diffraction. *Faraday Trans.* **1998,** *94* (17), 2615–2622.

[116] Campbell, J.W.; Hao, Q.; Harding, M.M.; Nguti, N.D.; Wilkinson, C. *LAUEGEN* version 6.0 and INTLDM. *J. Appl. Cryst.* **1998,** *31,* 496–502.

[117] Helliwell, J.R.; Wilkinson, C. X-ray and Neutron Laue Diffraction: Chaper XII in the Hercules, Vol. III, Neutron and Synchrotron Radiation for Condensed Matter Studies: Applications to Soft Condensed Matter and Biology; Springer Verlag Berlin, Heidelberg 1994.

[118] Deacon, A.; Gleichmann, T.; Kalb (Gilboa), A.J.; Price, H.; Raftery, J.; Bradbrook, G.; Yariv, J.; Helliwell, J.R. The Structure of Concanavalin A and its Bound Solvent Determined with Small-molecule Accuracy at 0.94Å Resolution. *Faraday Trans.* **1997,** *93* (24), 4305–4312.

[119] Habash, J.; Raftery, J.; Weisgerber, S.; Cassetta, A.; Lehmann, M.; Hoghoj, P.; Wilkinson, C.; Campbell, J.W.; Helliwell, J.R. Neutron Laue Diffraction Study of Concanavalin A: The Proton of Asp28. *Faraday Trans.* **1997,** *93* (24), 4313–4317.

[120] Habash, J.; Raftery, J.; Nuttall, R.; Price, H.J.; Wilkinson, C.; Kalb, A.J.; Helliwell, J.R. Direct Determination of the Positions of the Deuterium Atoms of the Bound Water in Concanavalin A by Neutron Laue Crystallography. *Acta Crystallogr. Sect. D-Biol. Crystallogr.* **2000,** *D56,* 541–550.

[121] Blakeley, M.P.; Kalb (Gilboa), A.J.; Helliwell, J.R.; Myles, D.A.A. The 15K Neutron Structure of Saccharide Free Concanavalin A. *Proc. Natl Acad. Sci. USA* **2004,** *101,* 16405–16410.

[122] Gilboa, A.J.K.; Myles, D.A.A.; Habash, J.; Raftery, J.; Helliwell, J.R. Neutron Laue Diffraction Experiments on a Large Unit Cell: Concanavalin A Complexed with Methyl-alpha-D-glucopyranoside. *J. Appl. Cryst.* **2001,** *34,* 454–457.

[123] Fisher, S.J.; Wilkinson, J.; Henchman, R.H.; Helliwell, J.R. An Evaluation Review of the Prediction of Protonation States in Proteins versus Crystallographic Experiment. *Crystallogr. Rev.* **2009,** *15* (4), 231–259.

[124] Fisher, S.J.; Helliwell, J.R. An Investigation into Structural Changes due to Deuteration. *Acta Cryst.* **2008,** *A64,* 359–367.

[125] Blakeley, M. Neutron Macromolecular Crystallography. *Crystallogr. Rev.* **2009,** *15,* 157–218.

[126] Wilkinson, C.; Lehmann, M.S.; Meilleur, F.; Blakeley, M.P.; Myles, D.A.A.; Vogelmeier, S.; Thoms, M.; Walsh, M.; McIntyre, G.J. Characterization of Image Plates for Neutron Diffraction. *J. Appl. Cryst.* **2009,** *42,* 749–757.

[127] Stuhrmann, H.B. Unique Aspects of Neutron Scattering for the Study of Biological Systems. *Rep. Progr. Phys.* **2004,** *67,* 1073–1115.

[128] Duke, E.M.H.; Johnson, L.N. Macromolecular Crystallography at Synchrotron Radiation Sources: Current Status and Future Developments. *Proc. R. Soc. Lond.* **2011,** *466,* 3421–3452.

[129] Dauter, Z.; Jaskolski, M.; Wlodawer, A. Impact of Synchrotron Radiation on Macromolecular Crystallography: A Personal View. *J. Synch. Rad.* **2011,** *17,* 433–444.

[130] Helliwell, J.R. Synchrotron Radiation Instrumentation, Methods and Scientific Utilisation, Chapter 8.1. In *International Tables Volume F Crystallography of Biological Macromolecules;* Rossmann, M.G., Arnold, E., Eds.; IUCr Kluwer: Dordrecht, 2001; Vol. 2011, pp. 189–204.

[131] Helliwell, J.R. 2001 Synchrotron Radiation in the Life Sciences Section in the Encyclopaedia of Life Sciences Publ Nature McMillan: X-ray Diffraction at Synchrotron Light Sources (Oct 2011). In Encyclopedia of Life Sciences; Wiley & Sons: Chichester, 2011, http://www.els.net/ [DOI: 10.1002/9780470015902.a0003109.pub2].

[132] Huxley, H.E.; Holmes, K.C. Development of Synchrotron Radiation as a High-Intensity Source for X-ray Diffraction. *J. Synch. Rad.* **1997,** *4,* 366–379.

[133] Harmsen, A.; Leberman, R.; Schulz, G.E. Comparison of Protein Crystal Diffraction Patterns and Absolute Intensities from Synchrotron and Conventional X-ray Sources. *J. Mol. Biol.* **1976,** *104,* 311–314.

[134] Ealick, S.E.; Rule, S.A.; Carter, D.C.; Greenhough, T.J.; Babu, Y.S.; Cook, W.J.; Habash, J.; Helliwell, J.R.; Stoeckler, J.D.; Parks, R.E. Jr; Chen, S.; Bugg, C.E. Three-dimensional Structure of Human Erythrocytic Purine Nucleoside Phosphorylase at 3.2Å Resolution. *J. Biol. Chem.* **1990,** *265,* 1812–1820.

[135] Bugg, Charles E.; Carson, William M.; John, A. Montgomery Drugs by Design: Structure-based Design, An Innovative Approach to Developing Drugs has Recently Spawned Many Promising Therapeutic Agents, Including Several now in Human Trials for Treating AIDS, Cancer and other Diseases. *Sci. Am.* **1993,** *December,* 92–98.

[136] Helliwell, J.R.; Hughes, G.; Przybylski, M.M.; Ridley, P.A.; Sumner, I.; Bateman, J.E.; Connolly, J.R.; Stephenson, R. A 2-D MWPC Area Detector for Use with Synchrotron X-radiation at the Daresbury Laboratory for Small Angle Diffraction and Scattering. *Nucl. Instrum. Methods.* **1982,** *201,* 175–180.

[137] Lewis, R. Multiwire Gas Proportional Counters: Decrepit Antiques or Classic Performers? *J. Synch. Rad.* **1994,** *1,* 43–53.

[138] Fourme, R. Position-sensitive Gas Detectors: MWPCs and their Gifted Descendants. *Nucl. Instrum. Methods Phys. Res. Sect. A: Acceler., Spectrom, Detect. Assoc. Equip.* **1997,** *392* (1–3), 1–11.

[139] Helliwell, J.R.; Ealick, S.; Doing, P.; Irving, T.; Szebenyi, M. Towards the Measurement of Ideal Data for Macromolecular Crystallography using Synchrotron Sources. *Acta Cryst.* **1993,** *D49,* 120–128.

[140] (a) Nielsen, F.S.; Lee, P.; Coppens, P. Crystallography at 0.3Å; Single-crystal Study of $Cr(NH_3)_6Cr(CN)_6$ at the Cornell High-energy Synchrotron Source. *Acta Cryst.* (1986). B42, 359–364; (b) P Coppens (with contributions from Cox, D., Vlieg, E., Robinson, I.K.). 1992 Synchrotron Radiation Crystallography; Academic Press, London. ISBN0-12-188080-X.

[141] Hart, M. Perfect Crystals in Crystal Structure Analysis. *Acta Cryst.* **1995,** *B51,* 483–485.

[142] Cianci, M.; Antonyuk, S.; Bliss, N.; Bailey, M.W.; Buffey, S.G.; Cheung, K.C.; Clarke, J.A.; Derbyshire, G.E.; Ellis, M.J.; Enderby, M.J.; Grant, A.F.; Holbourn, M.P.; Laundy, D.; Nave, C.; Ryder, R.; Stephenson, P.; Helliwell, J.R.; Hasnain, S.S. A high-throughput Structural Biology/Proteomics Beamline at the SRS on a New Multipole Wiggler. *J. Synch. Rad.* **2005,** *12,* 455–466.

[143] Helliwell, J.R.; Bell, A.M.T.; Bryant, P.; Fisher, S.J.; Habash, J.; Helliwell, M.; Margiolaki, I.; Kaenket, S.; Watier, Y.; Wright, J.; Yalamanchilli, S. Time-dependent Analysis of K_2PtBr_6 Binding to Lysozyme Studied by Protein Powder and Single Crystal X-ray Analysis. *Z. Krist.* **2011,** *225,* 570–575.

[144] Gwyndaf Evans, Danny Axford, David Waterman & Robin L. Owen. Macromolecular Microcrystallography. *Crystallogr. Rev.* **2011,** *17,* 105–142.

[145] Paithankar, K.S.; Sørensen, H.O.; Wright, J.P.; Schmidt, S.; Poulsen, H.F.; Garman, E.F. Simultaneous X-ray Diffraction from Multiple Single Crystals of Macromolecules. *Acta Cryst.* **2011,** *D67,* 608–618.

[146] Margiolaki, I.; Wright, J.P. Powder Crystallography on Macromolecules. *Acta Cryst.* **2008,** *A64,* 169–180.

[147] Helliwell, M. Anomalous Scattering for Small-molecule Crystallography. *J. Synch. Rad.* **2000,** *7,* 139–147.

[148] Helliwell, M.; Helliwell, J.R.; Kaucic, V.; Zabukovec Logar, N.; Teat, S.J.; Warren, J.E.; Dodson, E.J. Determination of Zinc Incorporation in the Zn Substituted Gallophosphate ZnULM-5 by Multiple Wavelength Anomalous Dispersion Techniques. *Acta Cryst.* **2011,** *B66,* 345–357.

[149] Maclean, E.J.; Rizkallah, P.J.; Helliwell, J.R. Protein Crystallography and Synchrotron Radiation; Current Status and Future Landscape. *Eur. Pharma. Rev.* **2006,** *2,* 71–76.

[150] New Light on Science: The Social & Economic Impact of the Daresbury Synchrotron Radiation Source (1981–2008) 214 pp. Report available from the Science and Technology Facilities Council at www. stfc.ac.uk or with google use the search term "SRS social and economic impact 1981–2008" (accessed 2 August 2015).

[151] Helliwell, J.R. Protein Crystallographic Drug Design using Synchrotron X-radiation. *Acta Radiol.* **1983,** *365* (Suppl.), 35–37.

[152] Boggon, T.J.; Helliwell, J.R.; Judge, R.A.; Olczak, A.; Siddons, D.P.; Snell, E.H.; Stojanoff, V. Synchrotron X-ray Reciprocal-Space Mapping, Topography and Diffraction Resolution Studies of Macromolecular Crystal Quality. *Acta Crystallogr. Sect. D-Biol. Crystallogr.* **2000,** *D56,* 868–880.

[153] Helliwell, M.; Helliwell, J.R.; Hanson, J.C.; Ericsson, T.; Kvick, A.; Kaucic, V.; Frampton, C.; Cassetta, A. Anomalous Dispersion Analyses of the Ni-atom Site in an Aluminophosphate Test Crystal Including the Use of Tuned Synchrotron Radiation. *Acta Cryst. B* **1996,** *52,* 479–486.

[154] Helliwell, M.; Helliwell, J.R.; Kaucic, V.; Logar, N.Z.; Barba, L.; Busetto, E.; Lausi, A. Determination of the Site of Incorporation of Cobalt in CoZnPO–CZP by Multiple-wavelength Anomalous-dispersion Crystallography. *Acta Cryst.* **1999,** *B55,* 327–332.

[155] Helliwell, J.R. Macromolecular crystallography using Synchrotron Radiation. Progress with Station 9.5 and a Novel, Toast-rack Detector Scheme for Laue Diffraction. *Nucl. Instrum. Methods.* **1991,** *A308,* 260–266.

[156] Helliwell, J.R. Macromolecular Crystal Twinning, Lattice Disorders and Multiple Crystals. *Crystallogr. Rev.* **2008,** *14,* 189–250.

[157] Chayen, N.E.; Helliwell, J.R.; Snell, E.H. "Macromolecular Crystallization and Crystal Perfection". Oxford University Press International Union of Crystallography Monographs on Crystallography (2010) ISBN-10: 0199213259. Oxford.

[158] Sax Mason question to John R Helliwell at the end of the Lonsdale Lecture.

[159] Jiang, J.; Sweet, R.M. Protein Data Bank depositions from synchrotron sources. *J. Synch. Rad.* **2004,** *11,* 319–327.

[160] Holton, J.M. A Beginner's Guide to Radiation Damage. *J. Synch. Rad.* **2009,** *16,* 133–142.

[161] Garman, E.F.; McSweeney, S.M. Progress in Research Into Radiation Damage in Cryo-cooled Macromolecular Crystals. *J. Synch. Rad.* **2007,** *14,* 1–3.

[162] Rhys, N.H.; Wang, M.-.C.; Jowitt, T.A.; Helliwell, J.R.; Grossmann, J.G.; Baldock, C. Deriving the Ultrastructure of α-crustacyanin using Lower-resolution Structural and Biophysical Methods. *J. Synch. Rad.* **2011,** *18,* 79–83.

[163] Cruickshank, D.W.J.; Helliwell, J.R.; Johnson, L.N. Eds. "Time–Resolved Macromolecular Crystallography" Proceedings of a Royal Society Discussion Meeting (1992) published by Oxford University Press. Oxford.

[164] Helliwell, J.R.; Rentzepis, P.M. Eds. "Time–Resolved Diffraction" Oxford University Press 1997. Oxford.

[165] Helliwell, J.R. New Opportunities in Biological and Chemical Crystallography. *J. Synch. Rad.* **2002,** *9,* 1–8.

(c)

FIGURE 1.1 The University of Manchester School of Chemistry Open Days included a crystallography display. (c) shows a different view of such an exhibit; for details and names of staff involved see text.

(b)

FIGURE 1.2 My Royal Institution ('The RI') Friday Evening Discourse April 2004. 'Why does a lobster change colour on cooking?' (b) With the RI Director Baroness Susan Greenfield and Dr Peter Zagalsky, great expert on marine colouration biochemistry. (Note the Director's carefully chosen colour of her evening dress!)

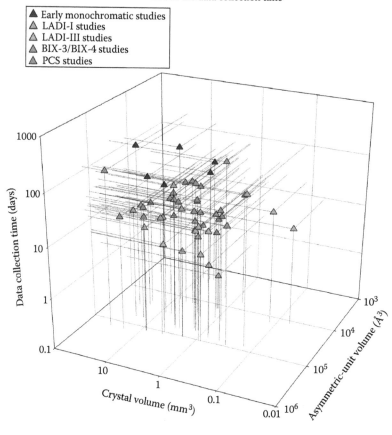

FIGURE 4.14 Three-dimensional scatter plot of the asymmetric unit-cell volume vs. crystal volume vs. data collection time for the various neutron structures solved so far. The new instruments are able to collect data from larger asymmetric unit cells, from smaller crystal volumes and with shorter data collection times. The original necessity to have crystal volumes of several mm³ is no longer the case with data now being collected from perdeuterated protein crystals close to 0.1 mm³. (Includes unpublished results from LADI-III.) From ref. (*125*) with permission of the author and Informa UK.

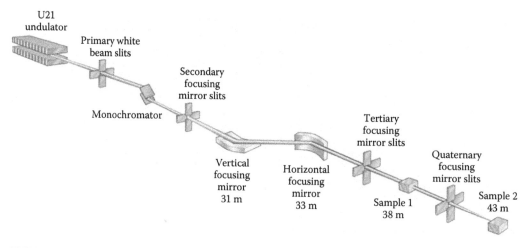

FIGURE 4.15 The layout of the DLS I19 chemical crystallography beamline. With permission of DLS and Professor Dave Allan, Scientist in charge.

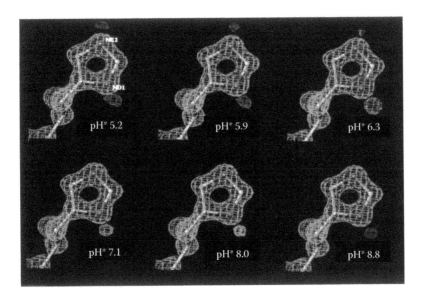

FIGURE 5.10 Sequence of snapshots showing the deprotonation of the imidazole ring of the His12 Nε2 atom: the difference maps are represented with a pH-scale colour code from red/acid to blue/basic (Reprinted from *J. Mol. Biol.,* 292, Berisio, R.; Lamzin, V.S.; Sica, F.; Wilson, K.S.; Zagari, A.; Mazzarella, L., Protein titration in the crystal state, Figure 5.3, © 1999, with permission from Elsevier).

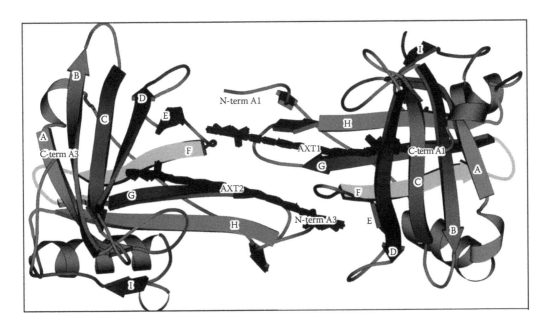

FIGURE 6.1 The structure of the A1/A3 dimer assembly (in ribbon format) with the two bound astaxanthin carotenoids (in 'stick' format labelled AXT1 and AXT2). The individual β strands which are shown in the colours red, green and blue are the amino acid sequence 'consensus regions' of the lipocalin protein family. Reproduced from (*3*), Copyright, 2002, National Academy of Sciences USA.

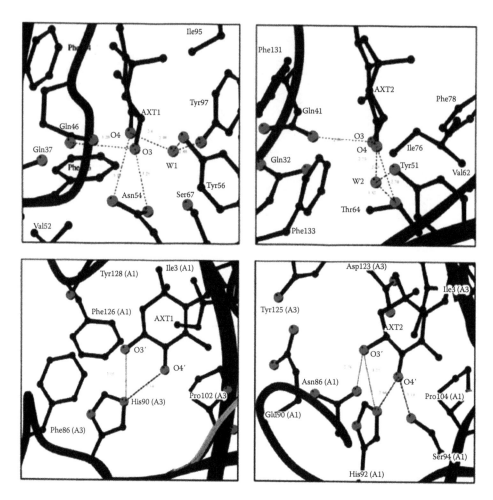

FIGURE 6.3 The detailed layout of the astaxanthin-binding sites at the end-ring molecular environments in β-crustacyanin (based on the crystal structure presented in Cianci et al. (*3*)). Top left: C1–6 end ring of AXT1 bound to the A1 molecule; top right: C1–6 end ring of AXT2 bound to the A3 molecule; bottom left: C1′–6′ end ring of AXT1 bound to the A3 molecule; bottom right: C1′–6′ end ring of AXT2 bound to the A1 molecule. Each astaxanthin molecule is equally shared between the two subunits A1 and A3. Copyright, 2002, National Academy of Sciences USA.

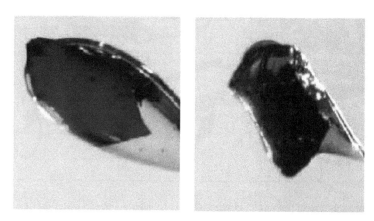

FIGURE 6.5 Crystals of the s-*cis* (left) and s-*trans* (right) conformers of astaxanthin diacetate (held in their loops for X-ray data collection). Reproduced from (*13*) with permission of the authors and the IUCr.

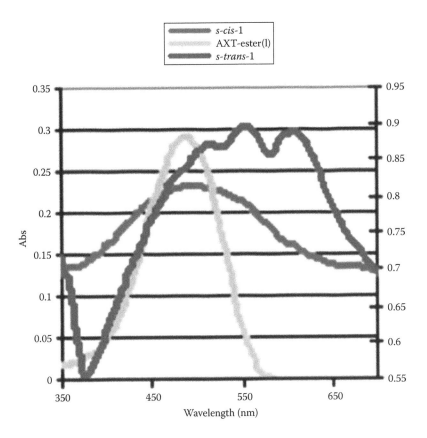

FIGURE 6.6 Solution-state (chloroform) UV-Vis spectrum of the diacetate ester of astaxanthin (green curve) and solid-state UV-Vis spectra of the s-*cis* and s-*trans* conformers, s-*cis*-**(1)** and s-*trans*-**(1)**. Reproduced from (*13*) with permission of the authors and the IUCr.

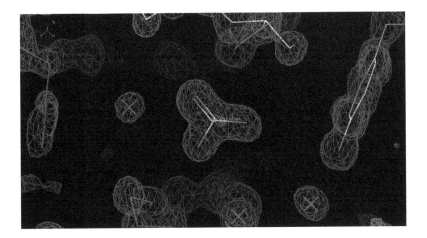

FIGURE 8.2 Discrete nitrate ion binding to a protein surface. In this molecular graphics representation, the blue mesh is the electron density derived from the X-ray diffraction from a single crystal of a protein (lysozyme in a triclinic crystal form). The interpretation of this in terms of chemistry then can readily be made: on the left, a nitrate ion and on the right, a tyrosine amino acid.

FIGURE 8.3 The UK's Science and Technology Facilities Council (STFC) undertook an analysis of the social and economic impact of the Daresbury Laboratory Synchrotron Radiation Source, which operated from 1981 to 2008; this figure shows the front cover of the report. The full report is available at www.stfc.ac.uk or with google use the search term "SRS social and economic impact 1981–2008" (accessed 2 August 2015).

5

An evaluation review of the prediction of protonation states in proteins *versus* crystallographic experiment*†

Stuart J. Fisher[a,b], James Wilkinson[a],
Richard H. Henchman[a,c], and John R. Helliwell[a‡]

*aSchool of Chemistry, The University of Manchester, Manchester M13 9PL, UK;
bInstitut Laue Langevin, 6 rue Jules Horowitz, BP 152, Grenoble, Cedex 9, 38042, France;
cManchester Interdisciplinary Biocentre, The University of Manchester,
131 Princess St, Manchester M1 7DN, UK*

(Received 3 June 2009; final version received 27 July 2009)

The known protonation states of protein crystal structures obtained using X-ray and neutron crystallographic data, and including relevant NMR-derived experimental information, have been predicted using three pK_a calculation tools, namely, PROPKA, H++ and MCCE. Comparisons between the experimental and predicted protonation states have been carried out in order to assess whether the results are of sufficient quality to validate their use in predicting the protonation states of two key histidine residues in the lobster *carapace* colouration protein β-crustacyanin as an example. Significant interest has been shown in the protonation states of these residues, which have been out of reach of experiment thus far and are likely to remain so.

Keywords: protonation states; protein pK_a predictions; amino acids; neutron crystallography; X-ray crystallography; NMR

Contents

* From *Crystallography Reviews,* Vol. 15, No. 4, October–December 2009, 231–259.

† Dedication: 2009 is the 50th Anniversary of the determination of the first protein crystal structure, myoglobin, by a team led by John Kendrew and which included Richard Dickerson and Bror Strandberg. Myoglobin crystal structures feature in our analyses here and so we wish to make our contribution to this important Anniversary by dedicating our article to that pioneering research.

‡ Email: john.helliwell@manchester.ac.uk

5.1 Introduction

Hydrogen atoms comprise 5–10% of a protein's mass and approximately 50% of the atoms by number, and the position of certain hydrogen atoms can play a major role in protein function. Whilst atomic positions can be determined using either X-ray or neutron diffraction, X-rays will predominantly give information about electron-rich, non-hydrogen atoms, whereas neutrons allow the position of hydrogen atoms to be determined. This is due to the fact that X-rays are scattered proportionally to the number of electrons; neutrons, however, are scattered by the atomic nucleus and as such, hydrogen and deuterium scatter with the same order of magnitude as carbon or oxygen (*1*). For example Kossiakoff and Spencer (*2*) used neutron protein crystallography to identify that His57 in trypsin was protonated (determined as deuterium as required by the method). Neutron protein crystallography typically between 1.5 and 2.7 Å resolution and ultra-high resolution X-ray structure analysis, typically around 0.9 Å, together allow the determination of the positions of deuterium/hydrogen atoms with very high confidence.

In this study, several proteins with known residue protonation states obtained using X-ray and neutron crystallographic data have been studied using three pK_a calculation tools. These tools are PROPKA, H++ and MCCE (see below for detailed references and websites). Another available tool is a custom version of the program WHAT IF (*3*) specifically for pK_a calculations (*4*); however, this is not freely available and so we do not consider this program further in this evaluation review. Predicted protonation states have been determined from the estimated pK_a and associated pH values. Comparisons of these predictions with results from neutron and X-ray crystallography as well as NMR have been carried out to assess if they are of sufficient quality to validate their use in predicting the protonation states of the potentially key histidine residues in the lobster *carapace* protein β-crustacyanin.

Our results can complement physical and theoretical chemistry methods, which might benefit from the improvement in the crystallographic experimental results exemplified here, thus leading to a better predictive ability. Furthermore, crystallographers interested in using ultra-high resolution X-ray and/or neutron protein crystallography will be interested in the quality of such predictions, notably of the ionizable residues Asp, Glu and His, which are the subject of an increasingly large number of protein studies with neutron macromolecular crystallography (for a recent review, see (5)).

Of course, each prediction tool used here has undergone careful self-validation tests (references below). Our study however not only includes a comparison of all three prediction tools but also focusses on an expanding crystallographic test set of proteins including larger molecular weight proteins and specifically the determination of protonation states instead of pK_a values.

Considerable effort is going into production of deuterated proteins and/or improved neutron sources and instruments in Europe, United States and Japan. Synchrotron X-ray beamlines are also being continually improved and new or enhanced sources worldwide are being developed, and applied to improved hydrogen and hydration state determination of proteins from crystallography. Crystallization techniques are also improving. Overall, it is important therefore to assess predictive tools and, not least, perhaps help improve them as a fruit of these experimental programmes.

5.2 Protein chemistry and potentially ionizable amino acids

A number of amino acid side chains in a protein are classically regarded as charged. These include aspartate, glutamate, histidine, lysine and arginine. Of these, lysine and arginine are generally not of interest, since they titrate outside the range of physiological pH values. Serine, threonine, tyrosine and cysteine can titrate at physiological pH values if they are located in special environments and hence are sometimes charged. In addition, the N- and C-termini of the polypeptide chain can titrate as well.

The three-dimensional position of a proton on a particular side chain can be interesting in determining the exact details of an enzyme mechanism. In the case of aspartate and glutamate, they have symmetrical carboxyl groups and so associated hydrogen bond donor and acceptors determine which oxygen is protonated in the folded protein state. In the case of lysine, however, this exact position is uninteresting, since all the protons are located on the same nitrogen atom.

Metal ligands in proteins strongly affect their immediate environment. They bind to deprotonated, charged, residues. Metal ions will, like other charged groups, affect the pK_a values of their neighbouring groups.

Standard pK_a values of the free amino acids are summarized in Table 5.1.

Table 5.1 Standard pK_a values for amino acids in solution

Residue	pK_a
Asp	3.9
Glu	4.1
Arg	12.5
Lys	10.5
His	6.0, 14.5
Cys	8.2
Tyr	10.5

Source: Reproduced from Liljas *et al.* (6), p. 14, Table 2.1.

5.3 pK_a values in proteins and calculation of estimated values

Proteins possess ionizable groups, which can be protonated or deprotonated depending on their pK_a values, and these are important for the overall protein stability. pK_a values of amino acid side chains play an important role in defining the pH-dependent characteristics of proteins (e.g. see (7)). When amino acids associate during protein folding, the titratable amino acids in the protein are relocated from a solution environment to an environment determined by the 3D structure of the protein and as such, the associated pK_a value is perturbed by interactions with the local environment.

5.3.1 *Experimental determination of* pK_a *values*

Nuclear Magnetic Resonance (NMR) spectroscopy is commonly used for determining pK_a values experimentally via titration. For a review, including such NMR studies applied to the well-known topic of the serine proteases, see Steitz & Shulman (8). In another example, transient hydrogen bonds were identified by solution NMR on the surface of hirudin (9). A very recent example is Søndergaard et al. (10) that includes pK_a calculations of protonation populations compared with NMR titration results. Furthermore, Powers and Jensen (11) have used predicted pK_a values of Asp and Glu residues whose 'PROPKA program's predictions are particularly accurate in the case of Asp and Glu residues' and thereby can be used to validate which structures in an NMR ensemble structure determination are the most accurate. Overall, NMR experimental methods and results have formed a platform for self-validation of the predictive tools. NMR spectroscopy is however limited to smaller molecular weight proteins due to the complex spectra obtained when studying larger proteins.

The protein pK_a database (PPD) (http://www.ddg-pharmfac.net/ppd/PPD/pKahomepage.htm) is a compilation of protein pK_a values, derived, for example using NMR and other spectroscopic techniques but thus far excluding crystallography. The PPD's aim is to serve as a standard for benchmarking of calculational methods and strategies. The PPD data were sourced from the primary literature and contains in excess of 1400 entries based on 157 proteins, including lysozyme and myoglobin which are reported here. The PPD database contains pK_a values for amino acid side chains, as well as the N and C termini. Over 75% of the entries focus on glutamate, lysine, histidine and aspartate. These four residues are all key ionizable residues, and therefore, the preponderance of these in the PPD is not driven by the PPD's selection, but by the available experimental data. For example very little data are currently available for arginine due to its pK_a value (12.5) which essentially precludes measurement by titration, as proteins will denature at highly basic pH. Clearly, the PPD is an important database. Since the pK_a estimation tools take input in the form of PDB coordinate files, often from crystallographic experiment, our results here provide a further significant cross-check between experimental and predicted results, which complement the NMR experimental results available in the PPD.

5.3.2 *Methods for calculating estimated values*

The mathematical methods that exist to calculate estimated pK_a values calculate the pK_a shift from the free energy of the protonated and unprotonated structures with an implicit solvent model (12–18), although useful results may still be obtained without any solvent term (19). Perturbation methods, such as thermodynamic integration, can evaluate the pK_a shift in the presence of explicit solvent (20). Calculating the pK_a directly requires more demanding methods that can model the electronic structure, such as quantum mechanical/molecular mechanical (21) or Car Parrinello (22) methods. At the other extreme are the empirical methods that fit to known pK_a values (7).

Much effort has recently been put into accounting for conformational flexibility and the coupling between multiple titratable residues (*13–15, 23*). Nielsen *et al.* (*4*) have shown the importance of optimizing the hydrogen bond network prior to electrostatic calculations and furthermore showed that correcting flipped Asn, His and Gln side chains can lead to changes in calculated pK_a values of up to 2.0 units.

5.3.3 *Tools for calculating estimated pK_a values*

5.3.3.1 *PROPKA*

PROPKA is a FORTRAN-based, web-accessible, program, which uses a set of empirical rules relating the protein structure to the pK_a values of ionizable residues for rapid predictions of pK_a values. This method was demonstrated by the developers to predict pK_a values with an overall root mean square (RMS) deviation of 0.79 pH units based on calculations for five proteins, 'OMTKY3', bovine pancreatic trypsin inhibitor, hen egg white lysozyme, RNase A and RNase H.

PROPKA uses three different kinds of pK_a perturbations: desolvation, hydrogen bonding and charge–charge interactions. Hydrogen bonding is the most common source of pK_a perturbations for Asp and Glu residues, while charge–charge interactions contribute only in about 10% of cases. Catalytic residues however appear to fall in this 10% of cases (*7*). Currently, the PROPKA program ignores possible pK_a shifts due to bound ligands, ions, or water molecules. The PROPKA program is freely available to the scientific community through a web interface at https://github.com/jensengroup/propka-3.1.

5.3.3.2 *The H++ web server*

The H++ server provides a tool that uses the finite difference Poisson–Boltzmann method, which is a modification of Poisson's equation that incorporates a description of the effect of solvent ions on the electrostatic field around a molecule. H++ computes pK_a values of amino acid side chains as well as other related characteristics of biomolecules, such as isoelectric points, titration curves and energies of protonation microstates. Protons are added to the user's input protein 3D structure according to the calculated ionization states of its titratable groups at the user-specified pH. This approach is different to the PROPKA program, which does not allow a user-specified pH to be input into the calculation (*12*). The H++ web-server is freely available to the scientific community through a web interface at http://biophysics.cs.vt.edu/H++.

5.3.3.3 *Multi-conformer continuum electrostatics*

Multi-conformer continuum electrostatics (MCCE) also uses the Poisson–Boltzmann method; however, unlike PROPKA, it allows for multiple conformations of hydroxyl and water protons, side chain rotamers and ligands, in the calculation of the pH dependence of the ionization equilibria of titratable groups (*13, 24*). The program's computations are broken down into three stages: (1) selection of conformers, (2) calculation of energy lookup tables and (3) Monte Carlo sampling and calculation of pK_as. The program outputs a calculated pK_a value for each ionizable residue as well as the stepwise values (pH 0–14 in steps of 1 unit) for the titration curves. MCCE is believed to provide better accuracy for prediction of histidine residues than the predictive tools discussed above (*25*). RMS deviations are of the order of 1 pH unit. The program has been previously tested using 166 residues in 12 proteins; the RMS error was 0.83 pH units, where 90% have errors <1 pH and only 3% have errors >2 pH (*13*). MCCE is freely available for download for academic usage at the following location: http://www.sci.ccny.cuny.edu/~mcce/.

5.3.4 *Prediction of protonation states and general points*

For each of the titratable residues of interest (histidine, aspartic acid and glutamic acid), a protonation state and an occupancy value have been calculated. The acid dissociation constant (K_a) of a
weak acid is defined as

$$K_a = \frac{\left[A^- \right]\left[H^+ \right]}{\left[HA \right]} \tag{5.1}$$

Taking logs of both sides and substituting for pH yields the Henderson–Hasselbalch equation which
relates the pH, the pK_a and the ratio of dissociated to associated acid:

$$pH = pK_a + \log\left(\frac{\left[A^- \right]}{\left[HA \right]} \right) \tag{5.2}$$

Rearrangement of the Henderson–Hasselbalch equation (Hasselbalch (*26*)) results in a percentage
of protonation for each residue at a given pH or an occupancy value, where, as a percentage:

$$\%_{[HA]} = \frac{100}{1 + \left(10^{pH - pK_a} \right)} \tag{5.3}$$

Each residue of interest has been examined and considered protonated if the $\%_{[HA]}$ is greater than 50%.

Using these pK_a prediction methods, it is only possible to determine whether a particular residue has
an attached proton or not, it is not possible to determine which particular atom the proton is attached
to or where it is physically located in three-dimensional space, i.e. in the case of the singly protonated
histidine (Figure 5.1), which nitrogen atom is concerned and for aspartic acid (Figure 5.2), which oxygen atom (glutamic acid not shown in the figure explicitly, since it also has a carboxyl side chain). The
chemical diagrams for the cases of lysine (Figure 5.3) and arginine (Figure 5.4) are also shown.

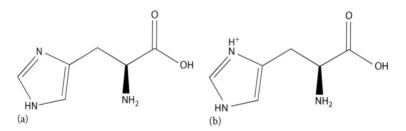

Figure 5.1 Chemical diagrams of (a) a singly protonated and (b) a doubly protonated histidine amino acid.

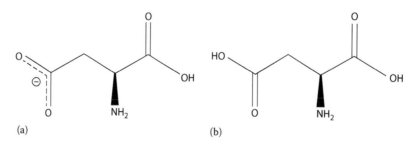

Figure 5.2 Chemical diagrams of (a) deprotonated and (b) protonated aspartic acid amino acid.

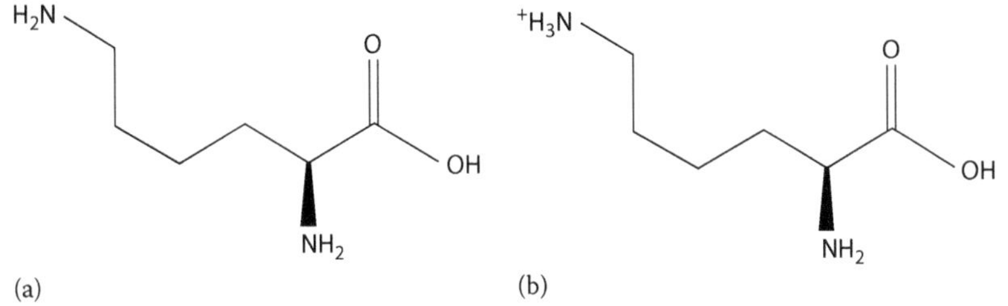

Figure 5.3 Chemical diagrams of (a) deprotonated and (b) protonated lysine amino acid.

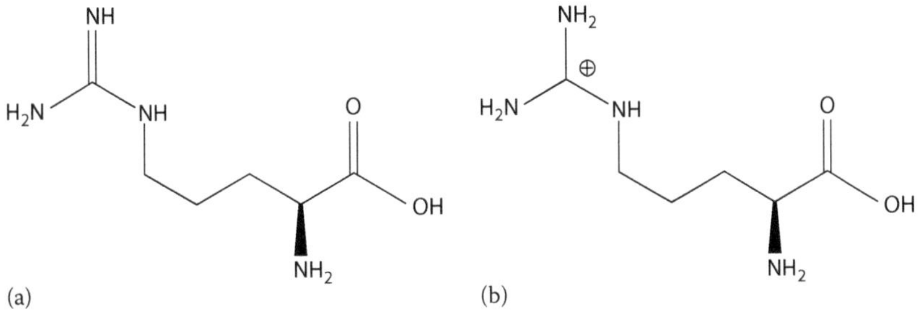

Figure 5.4 Chemical diagrams of (a) deprotonated and (b) protonated arginine amino acid.

5.4 Protein crystal structure test set

Proteins for which the hydrogen positions have been determined experimentally (as deuterium) to a very high confidence level by using neutron protein crystallography are especially favourable for inclusion in this study. Within the protein data bank (PDB), a number of such neutron protein structures have been identified. The file format is important as there is no consistent standard for labelling hydrogen atoms in a PDB file and MCCE is unable to process files that do not conform to its own internal standard, rendering it unable to process the neutron PDB input files for all four proteins: these were 2YZ4 for concanavalin A, 2GVE for D-xylose isomerase, 1LZN for lysozyme and 1L2K for myoglobin. Therefore, in order to compare results between the protonation predictions and experimentally known states, a number of very-high-resolution X-ray structures have been selected alongside the neutron structures. In order to keep these selections as consistent as possible, only X-ray crystal structures with the same pH as the neutron crystal structure have been selected and furthermore, alpha carbon and all atom RMS deviations have been calculated. MCCE requires separate pre-calculations for bound ligands prior to pK_a calculation, and therefore, proteins with bound organic ligands, such as aldose reductase, have been excluded.

The crystal structures are all finally refined and validated protein crystal structures deposited with the PDB. The summary tables below give the diffraction resolutions and the R and, where available, R_{free} reliability factors. The spread of resolutions, from 0.65 Å up to 3.23 Å, is the main cause of the spread of R and R_{free} values in the X-ray cases. In one case (1XIB), an R_{free} value is not available. In the neutron protein crystal structure refinements, two refinement protocols are in current use: one where there is a joint X-ray and neutron data model refinement and in this case, the R factors tend to be lower than the other protocol where the X-ray 'heavy atom' (i.e. carbon, nitrogen, oxygen in this context) positions are fixed at X-ray refinement positions and the neutron data are finally used to determine and refine deuterium atom positions. The quoted R and R_{free} for the latter protocol are the final neutron data refinement values, and since neutron data are weaker in intensity than X-ray data, the R and R_{free} values are higher than X-ray values at equivalent diffraction resolutions.

5.4.1 *Concanavalin A*

Concanavalin A is a 237-amino-acid lectin from the jack bean (*Canavalia ensiformis*), see Kalb (Gilboa) & Helliwell (*27*) for a review. It is by far one of the most thoroughly studied members of the legume lectin family of proteins (Weis & Drickamer (*28*)). It is a saccharide-binding protein although with no certain biological function (Deacon *et al.* (*29*)). The structure of the concanavalin A monomer is dominated by an extensive β-sheet strand arrangement, the so-called jellyroll motif, which associates as a dimer of dimers to form a tetramer of 100 kDa, and has been the study of numerous crystallographic researchers from the 1970s onwards and with crystals first grown even decades earlier by Sumner (*30*). The protein has been studied at 'chemical crystallography precision' and contains two metals, namely, a transition metal ion and a calcium ion, both necessary for saccharide binding (Deacon *et al.* (*29*)).

5.4.2 D-*Xylose isomerase*

D-Xylose isomerase (D-XI) is an enzyme from the bacterium *Streptomyces rubiginous*, which catalyzes the interconversion of the aldo-sugars D-xylose or D-glucose to the keto-sugars D-xylulose and D-fructose, respectively. It is a homotetramer of molecular mass 173 kDa that requires two divalent metal ions for function. The fold of its backbone, that of a $(\beta\alpha)_8$ barrel, was first determined in 1984 by Carrell *et al.* (*31*). One metal ion (M1) binds four carboxylate groups (Glu181, Glu217, Asp245 and Asp287). The other metal ion (M2) binds three carboxylate groups (Glu217, Asp257 and bidentate Asp255), and a His residue (His220). The carboxylate group of Glu217 is shared by both metal ions (Katz *et al.* (*32*)).

5.4.3 *Triclinic hen egg-white lysozyme*

Lysozyme is a 129 amino acid residue enzyme with its catalytic activity non-specifically targeted to bacterial cell walls and related to general non-specific organism defence. The main objective of the neutron-diffraction study (Bon *et al.* (*1*)) of triclinic hen egg-white lysozyme was to create a picture of the water molecules, but in addition, it was also found, as expected from various earlier studies, that the Glu35 residue was protonated and Asp52 was deprotonated at the crystallization pH of 4.7.

5.4.4 *Myoglobin*

Myoglobin is a 156 amino acid residue protein. The biological role of myoglobin as an oxygen storage protein is well-known, and will not be reiterated here. We highlight that Ostermann *et al.* (*33*) have deposited coordinates from a neutron protein crystallography study for this protein, and in addition, we have selected two very-high-resolution X-ray crystallography deposited coordinate sets (various are available and we apologise to anyone whose coordinates are not harnessed in this study).

5.5 Results

For each of the four proteins estimated, pK_a values have been calculated and a predicted protonation state inferred for each residue. The results tables show residues that are catalytically interesting or that have been highlighted in the respective publications for each protein. A full list of protonation states for every residue in each protein can be found in Figure 5.16 in Section 5.8. Likelihood statistics have been calculated for each program, for each residue type based on the complete set of residues (Table 5.10 in Section 5.5).

5.5.1 *Concanavalin A*

Three different concanavalin A structures were selected from the PDB: these were PDB code: 2YZ4, the neutron diffraction study by Ahmed *et al.* (*34*); 1NLS from the 0.94 Å resolution X-ray study by Deacon *et al.* (*29*) and 1JBC from the 1.15Å resolution X-ray study by Parkin *et al.* (*35*).

Protonation states of Asp82, His180, His24 and His51 were determined from the 2.2 Å neutron structure via direct inspection of the model file with hydrogen atoms (determined as deuteriums), while Asp28 and Glu102 are determined as protonated via bond length analysis (for the methods see Ahmed *et al.* (*34*)) of the 0.94 Å X-ray structure. Table 5.2 shows further details for these structure files.

From Figure 5.5, it can be seen that all three programs fail to predict the correct protonation state of Asp82, which is hydrogen bonded to a number of water molecules.

Of particular interest is that all three tools predict His24 as being doubly protonated when in fact, this residue has been found to be only singly protonated from the neutron study. This erroneous result is likely to be caused by the close proximity of the manganese metal ion. Other results of significance include His51, which is correctly predicted by PROPKA but not by H++ or MCCE, this residue appears to be hydrogen bonded to Thr194 which could influence the predictions. All three predictions tools correctly predict His180 as being doubly protonated in all three cases. This residue has four hydrogen-bonded water molecules, which are held in a cleft at one end of the molecule by Tyr54, Ser56, Leu81, Asp82, Ser113, His180, Ile181, Ala189 and Phe191 (*35*).

Table 5.2 Details for the three concanavalin A structures used for this study. RMS differences were calculated using the CCP4 program SUPERPOSE (Collaborative Computational Project 4 (*36*))

		2YZ4	1NLS	1JBC
Radiation probe		Neutron	X-ray	X-ray
Resolution (Å)		2.2	0.94	1.15
Reference		(*34*)	(*29*)	(*35*)
pH		6.5	6.8	6.5
RMS Cα (Å)		Reference structure	0.003[a]	0.286
RMS all (Å)		Reference structure	0.306	0.863
R factor (%)		28.1	13.2	14.2
R_{free} (%)		31.3	14.8	16.2

[a]The structure 2YZ4 was derived from 1NLS and all non-hydrogen atoms were fixed, hence the small RMS deviation.

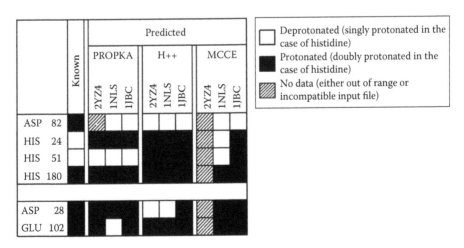

Figure 5.5 Graphical presentation of the predictions for the three concanavalin A coordinate sets by each of the three predictive tools are compared. The evidence for Asp28 and Glu102 being protonated most firmly stems from ultra-high resolution X-ray data bond distance analysis with the neutron evidence being weaker (*29, 34*).

Both PROPKA and MCCE also correctly predict Asp28 as being protonated, but H++ fails to predict this in all but the case of 1JBC. Glu102 is correctly predicted as being protonated by all three programs except the case of 1NLS for PROPKA.

All three programs successfully predict the majority of the deprotonated Asp and Glu residues with likelihoods of 83–94%. The programs are less accurate at predicting the protonated states of AspH with likelihoods of 17–50% (Table 5.3). The case for GluH however is better with a likelihood of 67–100%. In this particular case, the programs are better at predicting the doubly protonated state of HisH+ compared with that of His with likelihoods of 100% and 30–73%, respectively.

5.5.2 D-Xylose isomerase

Four different D-xylose isomerase structures were selected from the PDB: these were 2GVE, the neutron study by Katz *et al.* (*32*); 2GLK from the 0.94 Å resolution X-ray study by Katz *et al.* (*32*); 2GUB from the 1.80 Å resolution X-ray study by Katz *et al.* (*32*) and 1XIB from the 1.60 Å resolution X-ray study by Carrell *et al.* (*37*). Table 5.4 shows further details for the selected structure files.

From the neutron study of D-xylose isomerase (32, PDB code: 2GVE), it was seen that all the aspartic and glutamic acid side chains, at the pH of measurement (crystallization at pH 8.0), were deprotonated. All 10 histidine residues present within the protein were examined and placed into two groups, singly (His49, 71, 96, 243) or doubly protonated (His54, 198, 220, 230, 285, and 382).

Figure 5.6 shows the protonation prediction results for D-xylose isomerase. It can be seen that all three programs correctly predict all four singly protonated histidine residues in all four cases.

All three tools struggle at predicting doubly protonated histidine residues. H++ and MCCE appear more likely to predict when a His residue is doubly protonated as can be seen from the

Table 5.3 Likelihood of the three prediction programs correctly predicting the six residue types for concanavalin A

	No. of residues	Total	PROPKA (%)	H++ (%)	Total	MCCE (%)
Asp⁻	16	48	85	92	32	94
AspH	2	6	50	17	4	50
Glu⁻	6	18	89	83	12	83
GluH	1	3	67	100	2	100
His	5	15	73	33	10	30
HisH⁺	1	3	100	100	2	100

Table 5.4 Details for the four D-xylose isomerase structures used for this study

	2GVE	1XIB	2GLK	2GUB
Radiation probe	Neutron	X-ray	X-ray	X-ray
Resolution (Å)	1.8	1.60	0.94	1.80
Reference	(*32*)	(*37*)	(*32*)	(*32*)
pH	8.0	8.0	8.0	8.0
RMS Cα (Å)	Reference structure	0.167	0.258	0.181
RMS all (Å)	Reference structure	0.419	0.793	0.670
R factor (%)	27.1	15.2	11.5	16.0
R_{free} (%)	31.9	N/A	12.8	19.3

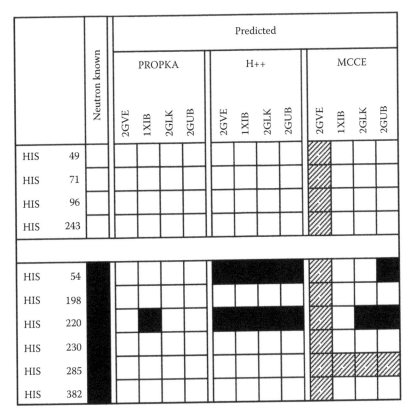

Figure 5.6 Graphical presentation of the His predictions for the four D-xylose isomerase coordinate sets by each of the three predictive tools compared.

results of His54 and His220. H++ predicts the correct results for all four of the PDB files for these two residues. Even so these programs still only manage to predict the doubly protonated residues in two of the six cases.

His220 is adjacent to one of the metal ions (M2) and as such is interesting in that H++ takes into account binding sites, whereas PROPKA ignores nearby water, ions and ligands, and therefore, it would be expected that H++ would be more accurate at predicting residues in these situations compared with PROPKA. This proximity of the metal ion means His220 would probably prefer not to be positively charged; it is however not pointing directly at the metal ion. PROPKA does however manage to predict the doubly protonated nature of His220 in one of the four cases. His54 is similar in that it is held in a rigid orientation by hydrogen bonds to a series of water molecules at the end of the active site near M2 (*37*) and so, H++ would be more likely to correctly predict the doubly protonated state of this residue.

It is interesting to note that the residues Glu181, 217, Asp245, 257 and 287 are all involved in metal binding with the M1 or M2 metal ions and, although the programs should predict these residues are deprotonated, they do in fact in some cases predict them as being protonated, as shown in Figure 5.7. This again illustrates that metal binding is an important, disruptive factor in pK_a calculation.

Once again the three programs successfully predict the majority of the deprotonated Asp and Glu residues with likelihoods of 86–95% and 85–96%, respectively (Table 5.5). The programs also successfully predict the correct protonation state of all of the neutral His residues with a likelihood of 100%; however, the predictions of the double protonated cases fair much worse with likelihoods in the range 4–33%.

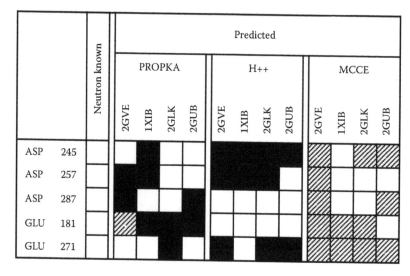

Figure 5.7 Graphical presentation of the Asp and Glu predictions for the four D-xylose isomerase coordinate sets by each of the three predictive tools compared.

Table 5.5 Likelihood of the three prediction programs correctly predicting the six residue types for D-xylose isomerase

	No. of residues	Total	PROPKA (%)	H++ (%)	Total	MCCE (%)
Asp⁻	37	148	86	95	111	89
AspH	0	–	–	–	–	–
Glu⁻	28	112	85	96	84	90
GluH	0	–	–	–	–	–
His	4	16	100	100	12	100
HisH⁺	6	24	4	33	18	17

5.5.3 Triclinic hen egg-white lysozyme

Three triclinic lysozyme structures were selected from the PDB; these were the quasi-Laue neutron diffraction study by Bon *et al.* (*1*) (PDB code: 1LZN), the 0.65 Å X-ray structure by Wang *et al.* (*38*) (PDB code: 2VB1) and the 0.93 Å X-ray structure Walsh *et al.* (*39*) (PDB code: 3LZT). Table 5.6 shows further details for the selected structure files. A search of the PPD database of Glu and Asp reveals a pK_a of 6.2 for Glu35 and 3.7 for Asp52 in chicken lysozyme from 1H NMR spectroscopy (*40*).

Table 5.6 Details for the three lysozyme structures used for this study

	1LZN	2VB1	3LZT
Radiation probe	Neutron	X-ray	X-ray
Resolution (Å)	1.7	0.65	0.93
Reference	(*1*)	(*38*)	(*39*)
pH	4.7	4.7	4.5
RMS C$_\alpha$ (Å)	Reference structure	0.304	0.296
RMS all (Å)	Reference structure	1.002	0.339
R factor (%)	20.4	8.48	9.3
R_{free} (%)	22.1	9.52	11.4

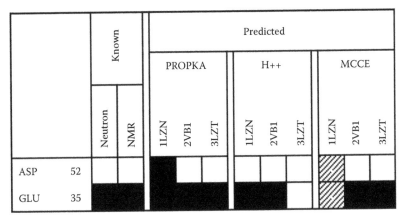

Figure 5.8 Graphical presentation of the predictions for the three lysozyme coordinate sets by each of the three predictive tools compared. NMR results are reproduced from Bartik *et al.* (*40*).

Figure 5.8 shows the protonation prediction results for lysozyme. In all cases except 3LZT for H++, the active site glutamic acid residue Glu35 is correctly predicted as being protonated. These results are consistent with both the neutron experiment and data available from NMR titrations. All three tools correctly predict the other catalytic residue, namely, Asp52 as being deprotonated in all but the case of PROPKA for 1LZN.

With lysozyme being a relatively small enzyme with only 129 residues, there are few other cases of interest with only 2 protonated glutamic residues acid (Glu7 and 35). Glu7 is only correctly predicted as being protonated by PROPKA in the case of 1LZN; no other tools in any of the cases manage to correctly predict this residue's protonation state. His15 although known to be deprotonated from neutron experiments is predicted as being doubly protonated in almost all cases except PROPKA for 1LZN. This residue could be affected by hydrogen bonding to nearby Thr89.

Again all three programs have a very high likelihood of predicting the deprotonated Asp residues with likelihoods in the range 95–100%. The likelihood of GluH being correctly predicted ranges between the three programs: for PROPKA it is 67%, for H++ 33% and for MCCE 50% (Table 5.7). All three programs struggle to correctly predict the neutral protonation state of His with likelihoods in the range of 0%–33%.

5.5.4 Myoglobin

Three myoglobin structures were selected from the PDB; these were the neutron study by Ostermann *et al.* (*33*) (PDB code: 1L2K), the 1.40 Å study by Arcovito *et al.* (*41*) (PDB code: 2JHO) and the 1.10 Å study by Vojtchovsky *et al.* (*42*) (PDB code: 1A6K). Table 5.8 shows further details for the selected structure files. A search on the PPD database reveals pK_a values for the histidine residues from 2D homonuclear double quantum NMR (*43, 44*) and 1D and 2D NMR (*45*). The Asp and Glu residues are all known to be deprotonated from the neutron study, and Figure 5.9 shows that all three

Table 5.7 Likelihood of the three prediction programs correctly predicting the six residue types for lysozyme

	No. of residues	Total	PROPKA (%)	H++ (%)	Total	MCCE (%)
Asp⁻	7	21	95	100	14	100
AspH	0	0	–	–	0	–
Glu⁻	0	0	–	–	0	–
GluH	2	6	67	33	4	50
His	1	3	33	0	2	0
HisH⁺	0	0	–	–	0	–

Table 5.8 Details for the three myoglobin structures used in this study

	1L2K	2JHO	1A6K
Radiation probe	Neutron	X-ray	X-ray
Resolution (Å)	1.5	1.40	1.10
Reference	*(33)*	*(41)*	*(42)*
pH	6.8	6.5	7.0
RMS C_α (Å)	Reference structure	0.447	0.222
RMS all (Å)	Reference structure	0.869	0.996
R factor (%)	20.1	18.0	13.1
R_{free} (%)	23.8	23.4	15.2

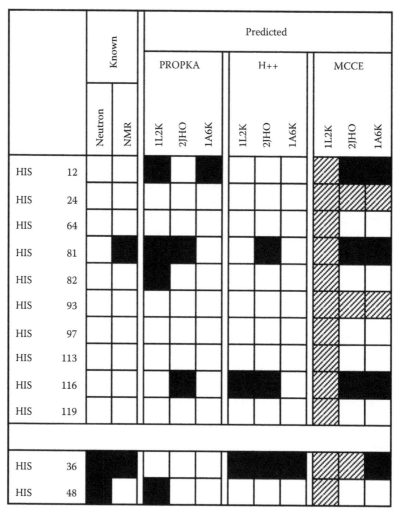

Figure 5.9 Graphical presentation of the predictions for the three myoglobin coordinate sets by each of the three predictive tools compared. NMR results are reproduced from Dalvit & Wright *(43)*; Ösapay *et al.* *(44)* and Cocco *et al.* *(45)*.

tools predict all of these residues as being deprotonated (see Figure 5.16 in Section 8) in all but the case of Glu59 for 1L2K using PROPKA.

There are however two doubly protonated histidine residues (His36 and 48). His36 is predicted (Figure 5.9) as being doubly protonated by both H++ and MCCE, but it is predicted as being singly protonated in all three cases by PROPKA. His48 is in close proximity to the sulphate-binding site, and as such, this may influence the accuracy of its prediction. All three programs predict this residue as being only singly protonated in all three cases except for PROPKA for the case of 1L2K.

There are also a number of singly protonated histidine residues (Figure 5.9): three of these are interesting, in that the prediction tools have predicted them in a number of cases as being doubly protonated. Of particular interest are His12, 81 and 116. Both His12 and 116 are hydrogen bonded to other residues: Gln128 in the case of His116 and Asp122 in the case of His12. His81 does not appear to directly interact with any other residues. Interestingly, the results from NMR titration and the neutron study differ in that the neutron study finds His48 to be protonated, whereas from the NMR titration, this residue has a pK_a of 5.3, which at the pH of crystallization would be singly and not doubly protonated. In this case, the prediction programs agree with the NMR titration result and not the neutron crystallography result. Also of interest is His81, which is known to be singly protonated from neutron studies; however, the NMR titration finds this residue to have a pK_a of 6.6, which at the pH of crystallization would make it doubly protonated. Again a number of the prediction programs agree with the NMR titration result instead of the neutron crystallography result. The risk of protein structure change due to deuteration is small, but not zero (*46*). We are not aware of the risks of NMR failing to produce a correct titration curve, but this may be our limited understanding of the NMR technique.

Once again all three programs successfully predict the deprotonated state of Asp and Glu residues in almost all cases with likelihoods in the range 98–100% (Table 5.9). For the case of the neutral His residues, the programs manage to predict this state in a number of cases: the likelihoods range from 50% to 90%.

5.5.5 *Overall statistics*

Table 5.10 shows the overall prediction likelihoods for each of the three pK_a calculation tools as well as the total overall likelihood for all three programs combined for the above protein test set. All three pK_a prediction tools show consistency at predicting the deprotonated charged states of both Asp and Glu residues. All three tools are relatively good at predicting the protonated state of Glu but less so for Asp with average likelihoods of 63% and 38%, respectively. It is clear however that the prediction of these two residues can be affected by proximal metal- and water-binding sites. Examples of this can be seen for D-xylose isomerase, in particular residues Glu181 and 217, and Asp245, 257 and 287.

Also all three programs in the majority predict singly protonated neutral His residues correctly as well with an average likelihood of 73%, but this being the default state for a histidine residue may not be the best test case. The case of doubly protonated His residues is more complex, and the three programs struggle to predict the double protonation of many of these residues. H++ and MCCE on average seem

Table 5.9 Likelihood of the three prediction programs correctly predicting the six residue types for myoglobin

	No. of residues	Total	PROPKA (%)	H++ (%)	Total	MCCE (%)
Asp⁻	7	21	100	100	14	100
AspH	0	0	–	–	0	–
Glu⁻	14	42	98	100	28	100
GluH	0	0	–	–	0	–
His	10	30	80	90	20	50
HisH⁺	2	6	17	50	4	25

Table 5.10 Overall likelihood of the three prediction programs correctly predicting the six residue types for all four proteins studied

	PROPKA		H++		MCCE		Overall	
	Total	%	Total	%	Total	%	Total	%
Asp⁻	238	88	238	95	171	92	647	92
AspH	6	50	6	17	4	50	16	38
Glu⁻	172	88	172	96	124	92	468	92
GluH	9	67	9	56	6	67	24	63
His	64	81	64	75	44	57	172	73
HisH⁺	33	15	33	42	24	25	90	28

to be more successful at this than PROPKA with likelihoods of 42% and 25%, respectively, *versus* 15%. There are a number of cases where the programs have correctly predicted doubly protonated histidines: these include His180 in concanavalin A, His54 and 220 in D-xylose isomerase. Again metal- and water-binding sites could be important for the correct prediction of these residues. His220 in D-xylose isomerase is nearby the binding site of metal M2 which could clearly influence the pK_a value.

We now proceed to evaluate two more cases armed with the statistics of Table 5.10. One case is ribonuclease A where very-high-resolution X-ray crystal structures at multiple pHs, a neutron macromolecular crystallography structure at an identical pH as one of the X-ray crystal structures and an NMR titration curve for the catalytic His12 are available. A second case is the crustacyanin His90 and His92, which are of wide interest to understand the basis of the bathochromic colour shift (or why a lobster changes colour on cooking).

5.6 Ribonuclease A: An example at multiple pH values

Pancreatic *bovine* ribonuclease A (RNase A) is a 124 amino acid residue endoribonuclease that cleaves and hydrolyzes single-stranded RNA. Due to the ready availability of protein material, RNase quickly became a protein standard test case. RNase A has been extensively studied using X-ray diffraction to atomic resolution (47) and neutron diffraction (48).

Both His12 and His119 are implicated in the catalytic mechanism. Berisio *et al.* (49) highlighted His12 (Figure 5.10) to study the protonation equilibrium present in the active site. His119 is found to be in multiple conformations at atomic resolution, and so it is unlikely that the hydrogen atoms would be visible on such a residue. Thus, His12 is the focus of Berisio *et al.* (49) and here.

Berisio *et al.* comment that:

The sequence of snapshots in Figure 5.10 shows the presence of hydrogen linked to the proximal Nδ1 atoms over the whole pH range. In contrast, hydrogen linked to Nε2 is present at acidic pH*, but disappears at higher pH* [pH* is the apparent proton activity of aqueous-organic solutions (50)]. This provides further experimental support for the proposed mechanism, which suggests that the Nε2 atom of His12 exchanges a proton with the substrate (see the review by Cuchillo *et al.* [(51)]). To our knowledge, this is the first direct structural evidence for the deprotonation of a residue in a crystalline environment. (49 reproduced with permission; material in square brackets added for clarification)

For a prediction test, seven ribonuclease A structures were selected from the PDB, the neutron study by Wlodawer & Sjolin (48), PDB code 5RSA, and the multiple pHs X-ray studies at atomic (1.05–1.15 Å) resolution (47, 49), with pH values and respective PDB references of: pH 5.2, 1KF2; pH 5.9, 1KF3; pH 6.3, 1KF4; pH 7.1, 1KF5; pH 8.0, 1KF7 and pH8.8, 1KF8. R_{free} factors were unavailable for each of the structures. Further details of the structures selected are shown in Table 5.11. A search of the PPD reveals pK_a values for His12 from 1H NMR spectroscopy (52).

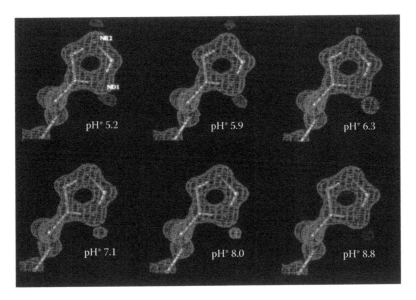

Figure 5.10 **(See colour insert.)** Sequence of snapshots showing the deprotonation of the imidazole ring of the His12 Ne2 atom: the difference maps are represented with a pH-scale colour code from red/acid to blue/basic (Reprinted from *J. Mol. Biol.*, 292, Berisio, R.; Lamzin, V.S.; Sica, F.; Wilson, K.S.; Zagari, A.; Mazzarella, L., Protein titration in the crystal state, Figure 5.3, © 1999, with permission from Elsevier).

Table 5.11 Details for the seven ribonuclease A structures used in this study

	5RSA	1KF2	1KF3	1KF4	1KF5	1KF7	1KF8
Radiation probe	Neutron	X-ray	X-ray	X-ray	X-ray	X-ray	X-ray
Resolution (Å)	2.0	1.10	1.05	1.10	1.15	1.15	1.15
Reference	(*48, 53*)	(*47*)					
pH	5.3	5.2	5.9	6.3	7.1	8.0	8.8
RMS Cα (Å)	Reference structure	0.149	0.146	0.145	0.177	0.195	0.203
RMS all (Å)	Reference structure	0.705	0.703	0.725	0.690	0.737	0.798
R factor (%)	18.3	10.3	10.2	10.4	10.6	10.6	10.6
R_{free} (%)	N/A	N/A	N/A	N/A	N/A	N/A	N/A

The results in Figure 5.11 show that both H++ and MCCE correctly predict His12 as being doubly protonated at pH5.3 (5RSA and 1KF2), which agrees with the neutron and NMR results. Furthermore H++ and MCCE both predict His12 to be protonated up to a pH of 5.9, which agrees well with the electron density maps in Figure 5.10.

Table 5.12 shows the actual calculated pK_a values from the three programs. It can be seen that in all cases, PROPKA predicts unusually small pK_a values for each of the His12 residues, even though histidine would be expected to titrate in the neutral range.

Closer inspection of the X-ray structures reveals a bound phosphate (in the case of the neutron structure) or sulphate (in the case of the X-ray structures) group adjacent to His12 and His119. This bound ligand could impact the calculated pK_a value. Unusually small pK_a values for histidine have been reported before (*54, 55*); for example Ruiz *et al.* (*55*) find for the case of aldose reductase that if the inhibitor is deleted (its volume is considered as a high dielectric region during the calculations), then the pK_a value for His110 is 0.7.

Further inspections of the relevant titration curves from H++ show some unusual behaviour. The titration curve for His12 in the neutron structure 5RSA does not show normal Henderson–Hasselbalch behaviour. It instead has a plateau region; Figure 5.12 shows this titration curve. Ondrechen *et al.*

	Known	Predicted																					
	Neutron / NMR	PROPKA							H++							MCCE							
		5RSA	1KF2	1KF3	1KF4	1KF5	1KF7	1KF8	5RSA	1KF2	1KF3	1KF4	1KF5	1KF7	1KF8	5RSA	1KF2	1KF3	1KF4	1KF5	1KF7	1KF8
HIS 12	■											■				▨						

Figure 5.11 Graphical presentation of the predictions for the seven ribonuclease A coordinate sets by each of the three predictive tools compared. NMR results are reproduced from Quirk & Raines (52).

Table 5.12 Estimated pKa values for His12 in ribonuclease A

	PROPKA	Occupancy (%)	H++	Occupancy (%)	MCCE	Occupancy (%)
5RSA	0.32	0	7.2	99	N/A	N/A
1KF2	2.18	0	6.2	91	8.52	100
1KF3	2.66	0	6.0	56	5.28	100
1KF4	2.41	0	6.0	33	4.90	9
1KF5	1.4	0	5.4	2	5.32	2
1KF7	1.44	0	5.1	0	5.11	0
1KF8	2.96	0	6.5	0	5.42	0

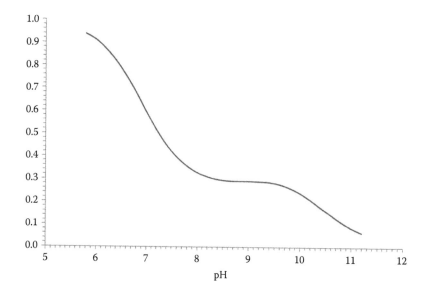

Figure 5.12 Calculated titration curve for His12 in 5RSA using H++.

(56) report that such a plateau region is characteristic of active site residues, and that by considering the titration curves of titratable residues in proteins, the active site can be identified.

5.7 Crustacyanin

α-crustacyanin is the colouration protein complex present in the lobster *carapace*, which has been implicated in the bathochromic shift and colour change from slate blue/black to orange/red that occurs upon cooking. The 350 kDa α-crustacyanin consists of an arrangement of β-crustacyanin subunits and the octameric unit of dimers consists of a mixture of five distinct chains A1, A2, A3,

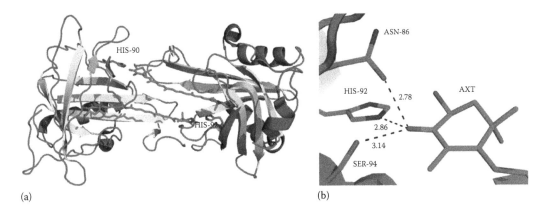

(a)   (b)

Figure 5.13 (a) Structure of beta-crustacyanin showing the two bound AXT molecules between the two dimers. (b) Distance between the His 92 residue and one of the AXT molecules in beta-crustacyanin showing that the keto-oxygens of the AXT are within hydrogen bonding distance of His 92. The same situation applies for the second AXT with respect to His 90. Figures based on (*58*) generated using PyMol (*59*).

C1 and C2. These can be classified into two groups exhibiting similar physical properties: type I gene sequences A1, C1 and C2 and type II gene sequences A2 and A3 (*57*). Determination of the X-ray structure of β-crustacyanin (*58*) enabled the identification of the structural features of the protein-bound chromophore astaxanthin (AXT) where β-crustacyanin exists as a heterodimer of A1 and A3 subunits with two AXT molecules bound between the two dimers positioned close to one another. Key interactions include those between His92 in the A1 subunit and His90 in the A3 subunit, each binding to one of the end rings of an AXT molecule (Figure 5.13).

Two main mechanisms for the bathochromic colour shift have been suggested. Firstly that the shift occurs due to polarization of the bound AXT molecules and secondly that it occurs due to an exciton interaction between the two AXT molecules. The polarization effect can occur via interactions of a nearby protonated residue with the keto end rings of the AXT molecule, in this case His90 and 92, respectively. The exciton interaction option can occur due to the close positioning of the two AXT molecules (*60*). Calculations by Durbeej & Eriksson (*61*) on AXT provide support that the bathochromic shift is due to the chromophore being polarized via hydrogen bonding interactions from a protonated histidine residue.

The β-crustacyanin (*58*) X-ray crystal structure does not determine directly the protonation states of the histidine residues due to the resolution (3.2 Å) of the crystal structure. Furthermore, the protein is likely to be out of reach of ultra-high resolution X-ray or neutron crystallography due to the large unit cell, high solvent content and small crystal size. The corresponding pK_a values have not been determined using NMR titration either. For these reasons, it is of interest to be able to predict protein protonation states of these histidine residues in particular. A double protonation prediction of these residues would further support the polarization hypothesis.

5.7.1 *Results*

There are six different PDB entries for the various subunits of α-crustacyanin (A1, C1 and C2) and one PDB entry for β-crustacyanin. Details for these seven crystal structures can be found in Table 5.13.

5.7.1.1 *β-crustacyanin*

The only available structure of β-crustacyanin is the X-ray diffraction study at 3.2 Å resolution ((*58*), PDB code: 1GKA). This structure was processed using the three prediction programs in order to evaluate the pK_a predictions of the two key His residues (His92 on chain A (A1) and His90 on

Table 5.13 Details for the seven crustacyanin structures used in this study

	1GKA	1I4U	1H91	1S44	1S2P	1OBQ	1OBU
Dimer subunits	A1/A3	C1/C1	A1/A1	C2/C2	C2/C2	C1/C1	C1/C1
pH.	9.0	9.0	9.0	9.0	7.0	9.0	9.0
Resolution (Å)	3.23	1.15	1.40	1.60	1.30	1.85	2.0
R factor (%)	21.33	15.0	17.66	20.3	18.9	16.9	17.4
R_{free} (%)	25.15	18.8	22.93	24.7	21.5	21.0	22.7

chain B (A3)) in β-crustacyanin. The protonation predictions (Figure 5.14) from all three prediction tools predict that both His90 and His92 are singly protonated.

5.7.1.2 Apo-crustacyanin

The apo-crustacyanin subunits have been studied at a much better X-ray resolution compared with that of β-crustacyanin. This results in improved accuracy of atom placement within the apo-crustacyanin structures if not yet experimental protonation states. The selected apo-crustacyanin structures all exist as homodimers with the key residue being His92. The six different PDB entries of the apo-crustacyanin subunits are 1I4U studied at an X-ray resolution of 1.15 Å by Gordon *et al.* (*62*), 1H91 studied at 1.40 Å by Cianci *et al.* (*63*), 1S44 studied at 1.60 Å and 1S2P studied at 1.30 Å by Habash *et al.* (*64*) and 1OBQ studied at 1.85 Å and 1OBU studied at 2.00 Å by Habash *et al.* (*65*).

The protonation predictions for the apo-crustacyanin structures (Figure 5.15) agree with the predictions for β-crustacyanin, in that none of the His92 residues are predicted to be doubly protonated. Since apo-crustacyanin is a homodimer, related by a non-crystallographic 2-fold symmetry axis in the crystal form studied, both His92 residues should be in chemically identical environments. Therefore, the predictions for both His92 residues should be very similar (Table 5.14) if not identical. H++ however shows large differences in pK_a values between the two chains. PROPKA shows the smallest deviations in pK_a values between the two chains, and in the case of 1H91, PROPKA actually predicts a pK_a of 1.24 for both residues. All of the MCCE pK_a values for each of the two chains are within one pK_a unit of each other, except the results for the PDB code: 1S44 where the deviation is 1.7 pK_a units.

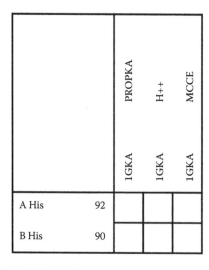

Figure 5.14 Graphical presentation of the predictions of the key His residues in β-crustacyanin for the PDB code: 1GKA (the experimental state of these two residues is not known from crystallography or NMR).

| | Predicted | | | | | | | | | | | | | | | | | | | | |
| | PROPKA | | | | | | | H++ | | | | | | | MCCE | | | | | | |
	1GKA	1I4U	1H91	1S44	1S2P	1OBQ	1OBU	1GKA	1I4U	1H91	1S44	1S2P	1OBQ	1OBU	1GKA	1I4U	1H91	1S44	1S2P	1OBQ	1OBU
A His 92																					
B His 92																					

Figure 5.15 Graphical presentation of the predictions of the key His residues in apo-crustacyanin.

Table 5.14 pK_a prediction values for the six apo-crustacyanin structures studied

	Residue	PROPKA	H++	MCCE
1U4U	Chain A His92	5.04	5.50	1.34
	Chain B His92	4.87	1.70	2.14
1H91	Chain A His92	1.24	−3.70	2.01
	Chain B His92	1.24	−4.00	1.70
1S44	Chain A His92	1.12	−4.50	1.22
	Chain B His92	1.09	−4.30	2.93
1S2P	Chain A His92	1.28	−4.40	2.35
	Chain B His92	1.31	−4.20	0.95
1OBQ	Chain A His92	5.30	4.60	2.00
	Chain B His92	4.97	0.40	2.50
1OBU	Chain A His92	5.38	4.70	2.76
	Chain B His92	5.21	1.20	OOR[a]

[a] Prediction out of range.

Many of the pK_a values for His92 in apo-crustacyanin are outside the normal range for histidine titration. In the case of the apo-crustacyanin structures, the region where the chromophore would be bound is replaced by a large solvent region and proximally bound sulphate ion. As previously mentioned, Ruiz *et al.* (*55*) see similar small pK_a values for a histidine residue in aldose reductase when the nearby inhibitor is removed. H++ has estimated very unusual negative pK_a values for His92 on both chains A and B.

5.7.2 Conclusions with regard to crustacyanin

From the predictions by all three tools for the His90 and 92 residues in β-crustacyanin, it suggests that both residues are only singly protonated. However, as described above for the test set of structures (summarized in Table 5.10), these three tools appear to have problems predicting when a residue is doubly protonated. Therefore, whilst our study does suggest that these key histidine residues are singly protonated, and thereby supporting the exciton theory put forward by Wijk *et al.* (*60*), it does not discount the theory (*61*) that the bathochromic shift could arise from the chromophore being polarized by a hydrogen-bonded doubly protonated histidine residue due to the low certainty in predicting doubly protonated His residues as evidenced by Table 5.10 statistics.

5.8 Overall conclusions

These investigations, summarized in Table 5.10 and Figure 5.16, confirm that experimental determination, such as from crystallography, remains paramount. Also, prediction methods do not predict which atom has the hydrogen atom in the case of a singly protonated histidine or a protonated

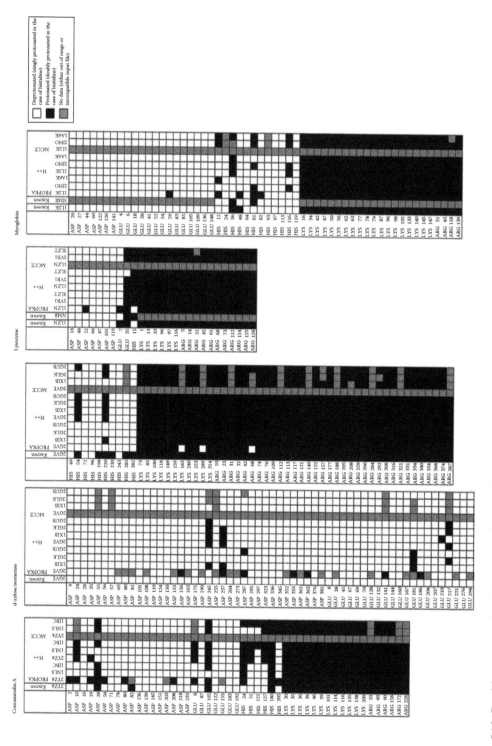

Figure 5.16 Graphic representation of the protonation states for each test protein for ALL residues (the main text has selected those residues that are functionally interesting or had received detailed comment in their respective publications).

carboxyl. Even for crystallography, though, there are risks and possible pitfalls. The use of crystallization chemical conditions may be at a pH remote from the functioning pH. Also kinetic and resonance effects of a key hydrogen atom may change upon deuteration, including pK_a shifts, thus affecting 3D structure. The prediction tools can add insight into all these effects. We have highlighted their predictive likelihoods with a test set of proteins, which were well determined from very-high-resolution and neutron macromolecular crystallography studies, along with experimental results from NMR in several cases. These proteins yielded a reasonably large number of Asp, Glu and His; the weakest part of the test set is the rather few AspH and GluH cases, but at least their protonation states were very well established. Armed with these statistical likelihoods we then, furthermore, evaluated them with ribonuclease A for which multiple atomic resolution X-ray crystal structures were available at six different pHs, highlighted by His12 and where neutron macromolecular crystallography and NMR results were known. Finally, the unknown, very interesting cases of crustacyanin His90 and His92 were predicted and for which are likely to remain out of range of experiment. The crustacyanin results are inconclusive, in that the statistical likelihoods of predictive ability tell us that doubly protonated histidines are the poorest in terms of predictive confidence. The value of this evaluation in this case is then to encourage improvements in the predictive methods themselves.

Acknowledgements

We are grateful to all the investigators who provided us with coordinates via the PDB, and who we have referenced here. For our own crystallographic investigations, we are grateful to the SR facilities at Cornell ('MacCHESS') for the 1NLS study during JRH's sabbatical stay in Cornell University, the SRS Daresbury (for all our crustacyanin studies) and the Institut Laue Langevin neutron facility LAue DIffractometer 'LADI' (now 'LADI III'). SJF gratefully acknowledges the PhD studentship support from the Institut Laue Langevin and the University of Manchester. We are especially grateful to the referees for their constructive suggestions that have clearly improved the presentation of the details of this wide-ranging study, i.e. encompassing multiple experimental and theoretical techniques.

Notes on the contributors

Stuart J. Fisher, after graduating from the University of Manchester in 2006 with a first class MChem Honours degree in Chemistry with Physics, is undertaking his PhD based at the Institut

Laue Langevin. He is supervised jointly by Professor John R. Helliwell (University of Manchester) and Dr Matthew P. Blakeley (instrument scientist responsible for the LAue DIffractometer 'LADI-III' at the Institut Laue-Langevin (ILL), Grenoble, France). His research interests include neutron and ultra-high resolution X-ray macromolecular crystallography methods and macromolecular structure and function including protonation states.

James Wilkinson studied at The University of Manchester from 2002 to 2007 and was awarded an MChem (Hons) in Chemistry in 2006 and an MSc in Cheminformatics in 2007.

Richard H. Henchman is a Lecturer in Theoretical Chemistry at the University of Manchester since 2005. He received his BSc (Hons) in Theoretical Chemistry at the University of Sydney in 1996 and his PhD in Physical Chemistry at the University of Southampton in 2000, and undertook postdoctoral research at the University of California, San Diego, from 2000 to 2004. He is currently working on methods to predict the structure of molecular systems.

John R. Helliwell is Professor of Structural Chemistry at the University of Manchester since 1989. He has longstanding research links with Daresbury Laboratory and is currently a scientific adviser to the STFC CLIK's Knowledge Transfer to Industry. He was awarded a BA and a DSc in Physics from the University of York in 1974 and 1996, respectively, and a DPhil from the University of Oxford in 1978 and is a Fellow of the Institute of Physics since 1986, of The Royal Society of Chemistry since 1995 and of the Institute of Biology since 1998. He is an Honorary Member of the National Institute of Chemistry, Ljubljana, Slovenia, since 1997. He is active in serving the wider crystallographic community, e.g. through the IUCr, the European Crystallographic Association and the British Crystallographic Association, as well as SR and neutron facilities via their advisory committees.

References

[1] Bon, C.; Lehmann, M.S.; Wilkinson, C. Quasi-Laue Neutron-Diffraction Study of the Water Arrangement in Crystals of Triclinic Hen Egg-Ehite Lysozyme. *Acta Crystallogr. D* **1999,** *55,* 978–987.

[2] Kossiakoff, A.A.; Spencer, S.A. Neutron Diffraction Identifies His 57 as the Catalytic Base in Trypsin. *Nature* **1980,** *288,* 414–416.

[3] Vriend, G. WHAT IF: A Molecular Modeling and Drug Design Program. *J. Mol. Graph. Model.* **1990,** *8,* 52–56.

[4] Nielsen, J.E.; Andersen, K.V.; Honig, B.; Hooft, R.W.W.; Klebe, G.; Vriend, G.; Wade, R.C. Improving Macromolecular Electrostatics Calculations. *Protein Eng.* **1999,** *12,* 657–662.

[5] Blakeley, M.P. Neutron Macromolecular Crystallography. *Crystallogr. Rev.* **2009,** *15,* 157–218.

[6] Liljas, A.; Liljas, L.; Piskur, J.; Lindblom, G.; Nissen, P.; Kjeldgaard, M. *Textbook of Structural Biology*; Singapore: World Scientific, 2009; p. 14.

[7] Li, H.; Robertson, A.D.; Jensen, J.H. Very Fast Empirical Prediction and Rationalization of Protein pK_a Values. *Proteins: Structure, Function, Bioinform.* **2005,** *61,* 704–721.

[8] Steitz, T.A.; Shulman, R.G. Crystallographic and NMR Studies of the Serine Proteases. *Ann. Rev. Biophys. Bioeng.* **1982,** *11,* 419–444.

[9] Szyperski, T.; Antuch, W.; Schick, M.; Betz, A.; Stone, S.R.; Wuthrich, K. Transient Hydrogen Bonds Identified on the Surface of the NMR Solution Structure of Hirudin. *Biochemistry* **1994,** *33,* 9303–9310.

[10] Søndergaard, C.R.; McIntosh, L.P.; Pollastri, G.; Nielsen, J.E. Determination of Electrostatic Interaction Energies and Protonation State Populations in Enzyme Active Sites. *J. Mol. Biol.* **2008**, *376*, 269–287.

[11] Powers, N.; Jensen, J.H. Chemically Accurate Protein Structures: Validation of Protein NMR Structures by Comparison of Measured and Predicted pK_a Values. *J. Biomol. NMR* **2006**, *35*, 39–51.

[12] Gordon, J.C.; Myers, J.B.; Folta, T.; Shoja, V.; Heath, L.S.; Onufriev, A. H++: A Server for Estimating pK_a's and Adding Missing Hydrogens to Macromolecules. *Nucl. Acids Res.* **2005**, *33*, W368–371.

[13] Georgescu, R.E.; Alexov, E.G.; Gunner, M.R. Combining Conformational Flexibility and Continuum Electrostatics for Calculating pK_a's in Proteins. *Biophys. J.* **2002**, *83*, 1731–1748.

[14] Mongan, J.; Case, D.A. Biomolecular Simulations at Constant pH. *Curr. Opin. Struct. Biol.* **2005**, *15*, 157–163.

[15] Li, X.; Jacobson, M.P.; Zhu, K.; Zhao, S.; Friesner, R.A. Assignment of Polar States for Protein Amino Acid Residues using an Interaction Cluster Decomposition Algorithm and its Application to High Resolution Protein Structure Modeling. *Proteins* **2007**, *66*, 824–837.

[16] Bashford, D. Macroscopic Electrostatic Models for Protonation States in Proteins. *Front. Biosci.* **2004**, *9*, 1082–1099.

[17] Kuhn, B.; Kollman, P.A.; Stahl, M. Prediction of pK_a Shifts in Proteins Using a Combination of Molecular Mechanical and Continuum Solvent Calculations. *J. Comput. Chem.* **2004**, *25*, 1865–1872.

[18] Liptak, M.D.; Shields, G.C. Accurate pK_a Calculations for Carboxylic Acids Using Complete Basis Set and Gaussian-n Models Combined with CPCM Continuum Solvation Methods. *J. Am. Chem. Soc.* **2001**, *123*, 7314–7319.

[19] Nielsen, J.E. Analysing the pH-Dependent Properties of Proteins Using pK_a Calculations. *J. Mol. Graph. Modell.* **2007**, *25*, 691–699.

[20] Simonson, T.; Carlsson, J.; Case, D.A. Proton Binding to Proteins: pK_a Calculations with Explicit and Implicit Solvent Models. *J. Am. Chem. Soc.* **2004**, *126*, 4167–4180.

[21] Shurki, A.; Warshel, A. Structure/Function Correlations of Proteins using MM, QM/MM, and Related Approaches: Methods, Concepts, Pitfalls, and Current Progress. *Adv. Prot. Chem.* **2003**, *66*, 249–313.

[22] Ivanov, I.; Chen., B.; Raugei, S.; Klein, M.L. Relative pK_a Values from First-Principles Molecular Dynamics: The Case of Histidine Deprotonation. *J. Phys. Chem. B* **2006**, *110*, 6365–6371.

[23] Warwicker, J. Improved pK_a Calculations through Flexibility Based Sampling of a Water-Dominated Interaction Scheme. *Prot. Sci.* **2004**, *13*, 2793–2805.

[24] Alexov, E.G.; Gunner, M.R. Incorporating Protein Conformational Flexibility into the Calculation of pH-Dependent Protein Properties. *Biophys. J.* **1997**, *72*, 2075–2093.

[25] Davies, M.N.; Toseland, C.P.; Moss, D.S.; Flower, D.R. Benchmarking pK_a Prediction. *BMC Biochemistry* **2006**, *7*, 18.

[26] Hasselbalch, K.A. Die Berechnung der Wassersroffzahl des Blutes ous der freien und gebunden Kohlensaure desselben, und die Sauerstoffbindung des Blutes als Funktion der Wasserstoffzahl. *Biochem. Z.* **1917**, *78*, 112–144.

[27] Kalb (Gilboa), A.J.; Helliwell, J.R. Concanavalin A. In *Handbook of Metalloproteins*; Messerschmidt, A., Huber, R., Poulos, T., Weghardt, K., Eds.; New York: John Wiley, 2001; pp. 963–372.

[28] Weis, W.I.; Drickamer, K. Structural Basis of Lectin-Carbohydrate Recognition. *Ann. Rev. Biochem.* **1996**, *65*, 441–473.

[29] Deacon, A.; Gleichmann, T.; Kalb (Gilboa), A.J.; Price, H.; Raftery, J.; Bradbrook, G.; Yariv, J.; Helliwell, J.R. The Structure of Concanavalin A and its Bound Solvent Determined with Small-Molecule Accuracy at 0.94 Å Resolution. *J. Chem. Soc., Faraday Trans.* **1997**, *93*, 4305–4312.

[30] Sumner, J.B. The Globulins of the Jack Bean, Canavalia ensiformis. *J. Biol. Chem.* **1919**, *37*, 137–141.

[31] Carrell, H.L.; Rubin, B.H.; Hurley, T.J.; Glusker, J.P. X-ray Crystal Structure of D-Xylose Isomerase at 4 Å Resolution. *J. Biol. Chem.* **1984**, *259*, 3230–3236.

[32] Katz, A.K.; Li, X.; Carrell, H.L.; Hanson, B.L.; Langan, P.; Coates, L.; Schoenborn, B.P.; Glusker, J.P.; Bunick, G.J. Locating Active-Site Hydrogen Atoms in D-Xylose Isomerase: Time-of-Flight Neutron Diffraction. *Proc. Natl Acad. Sci.* **2006**, *103*, 8342–8347.

[33] Ostermann, A.; Tanaka, I.; Engler, N.; Niimura, N.; Parak, F.G. Hydrogen and Deuterium in Myoglobin as seen by a Neutron Structure Determination at 1.5 Å Resolution. *Biophys. Chem.* **2002**, *95*, 183–193.

[34] Ahmed, H.U.; Blakeley, M.P.; Cianci, M.; Cruickshank, D.W.J.; Hubbard, J.A.; Helliwell, J.R. The Determination of Protonation States in Proteins. *Acta Crystallogr. D* **2007**, *63*, 906–922.

[35] Parkin, S.; Rupp, B.; Hope, H. Atomic Resolution Structure of Concanavalin A at 120 K. *Acta Crystallogr. D* **1996**, *52,* 1161–1168.

[36] Collaborative Computational Project 4 The CCP4 Suite: Programs for Protein Crystallography. *Acta Crystallogr. D* **1994**, *50,* 760–763.

[37] Carrell, H.L.; Hoier, H.; Glusker, J.P. Modes of Binding Substrates and their Analogues to the Enzyme D-Xylose Isomerase. *Acta Crystallogr. D* **1994**, *50,* 113–123.

[38] Wang, J.; Dauter, M.; Alkire, R.; Joachimiak, A.; Dauter, Z. Triclinic lysozyme at 0.65 Å Resolution. *Acta Crystallogr. D* **2007**, *63,* 1254–1268.

[39] Walsh, M.A.; Schneider, T.R.; Sieker, L.C.; Dauter, Z.; Lamzin, V.S.; Wilson, K.S. Refinement of Triclinic Hen Egg-White Lysozyme at Atomic Resolution. *Acta Crystallogr. D* **1998**, *54,* 522–546.

[40] Bartik, K.; Redfield, C.; Dobson, C.M. Measurement of the Individual pK_a Values of Acidic Residues of Hen and Turkey Lysozymes by Two-Dimensional 1H NMR. *Biophys. J.* **1994**, *66,* 1180–1184.

[41] Arcovito, A.; Benfatto, M.; Cianci, M.; Hasnain, S.S.; Nienhaus, K.; Nienhaus, G.U.; Savino, C.; Strange, R.W.; Vallone, B.; Della Longa, S. X-ray Structure Analysis of a Metalloprotein with Enhanced Active-site Resolution Using *in situ* X-Ray Absorption near Edge Structure Spectroscopy. *Proc. Natl Acad. Sci.* **2007**, *104,* 6211–6216.

[42] Vojtchovský, J.; Chu, K.; Berendzen, J.; Sweet, R.M.; Schlichting, I. Crystal Structures of Myoglobin-Ligand Complexes at Near-Atomic Resolution. *Biophys. J.* **1999**, *77,* 2153–2174.

[43] Dalvit, C.; Wright, P.E. Assignment of Resonances in the 1H Nuclear Magnetic Resonance Spectrum of the Carbon Monoxide Complex of Sperm Whale Myoglobin by Phase-Sensitive Two-Dimensional Techniques. *J. Mol. Biol.* **1987**, *194,* 313–327.

[44] Ösapay, K.; Theriault, Y.; Wright, P.E.; Case, D.A. Solution Structure of Carbonmonoxy Myoglobin Determined from Nuclear Magnetic Resonance Distance and Chemical Shift Constraints. *J. Mol. Biol.* **1994**, *244,* 183–197.

[45] Cocco, M.J.; Kao, Y.H.; Phillips, A.T.; Lecomte, J.T.J. Structural Comparison of Apomyoglobin and Metaquomyoglobin: pH Titration of Histidines by NMR Spectroscopy. *Biochemistry* **1992**, *31,* 6481–6491.

[46] Fisher, S.J.; Helliwell, J.R. An Investigation into Structural Changes due to Deuteration. *Acta Crystallogr. A* **2008**, *64,* 359–367.

[47] Berisio, R.; Sica, F.; Lamzin, V.S.; Wilson, K.S.; Zagari, A.; Mazzarella, L. Atomic resolution structures of ribonuclease A at six pH values. *Acta Crystallogr. D – Biol. Crystallogr.* **2002**, *58,* 441–450.

[48] Wlodawer, A.; Sjolin, L. Structure of Ribonuclease-A – Results of Joint Neutron and X-Ray Refinement at 2.0 Å Resolution. *Biochemistry* **1983**, *22,* 2720–2728.

[49] Berisio, R.; Lamzin, V.S.; Sica, F.; Wilson, K.S.; Zagari, A.; Mazzarella, L. Protein Titration in the Crystal State. *J. Mol. Biol.* **1999**, *292,* 845–854.

[50] Salomon, M. Thermodynamic measurements: Electrochemical measurements. In *Physical Chemistry of Organic Solvent Systems* (Covington, A. K. & Dickinson, T., Eds); Plenum Press: London and New York, 1973; pp. 137–219.

[51] Cuchillo, C.M.; Vilanova, M.; Nogues, M.V. Pancreatic ribonucleases. In *Ribonucleases: Structures and Functions* (editors: D'Alessio, G.; Riordan, J.F.); Academic Press: New York, 1997; pp. 271–300.

[52] Quirk, D.J.; Raines, R.T. His … Asp Catalytic Dyad of Ribonuclease A: Histidine pK_a Values in the Wild-Type, D121N, and D121A Enzymes. *Biophys. J.* **1999**, *76,* 1571–1579.

[53] Wlodawer, A.; Borkakoti, N.; Moss, D.S.; Howlin, B. Comparison of Two Independently Refined Models of Ribonuclease-A. *Acta Crystallogr. B* **1986**, *42,* 379–387.

[54] Varnai, P.; Warshel, A. Computer Simulation Studies of the Catalytic Mechanism of Human Aldose Reductase. *J. Am. Chem. Soc.* **2000**, *122,* 3849–3860.

[55] Ruiz, F.; Hazemann, I.; Mitschler, A.; Joachimiak, A.; Schneider, T.; Karplus, M.; Podjarny, A. The Crystallographic Structure of the Aldose Reductase-IDD552 Complex Shows Direct Proton Donation from Tyrosine 48. *Acta Crystallogr. D – Biol. Crystallogr.* **2004**, *60,* 1347–1354.

[56] Ondrechen, M.J.; Clifton, J.G.; Ringe, D. THEMATICS: A Simple Computational Predictor of Enzyme Function from Structure. *Proc. Natl Acad. Sci. USA* **2001**, *98,* 12473–12478.

[57] Quarmby, R.; Nordens, D.A.; Zagalsky, P.F.; Ceccaldi, H.J.; Daumas, R. Studies on the Quaternary Structure of the Lobster Exoskeleton Carotenoprotein, Crustacyanin. *Comparative Biochem. Physiol. B: Biochem. Mol. Biol.* **1977**, *56,* 55–61.

[58] Cianci, M.; Rizkallah, P.J.; Olczak, A.; Raftery, J.; Chayen, N.E.; Zagalsky, P.F.; Helliwell, J.R. The Molecular Basis of the Coloration Mechanism in Lobster Shell: Beta-Crustacyanin at 3.2 Å Resolution. *Proc. Natl Acad. Sci.* **2002,** *99,* 9795–9800.

[59] DeLano, W.L. The PyMOL Molecular Graphics System. *http://www.pymol.org* **2002.**

[60] Wijk, A.A.C.; Spaans, A.; Uzunbajakava, N.; Otto, C.; deGroot, H.J.M.; Lugtenburg, J.; Buda, F. Spectroscopy and Quantum Chemical Modeling Reveal a Predominant Contribution of Excitonic Interactions to the Bathochromic Shift in α-Crustacyanin, the Blue Carotenoprotein in the Carapace of the Lobster *Homarus gammarus. J. Am. Chem. Soc.* **2005,** 127, 1438–1445.

[61] Durbeej, B.; Eriksson, L.A. Protein-Bound Chromophores Astaxanthin and Phytochromoblilin: Excited State Quantum Chemical Studies. *Phys. Chem. Chem. Phys.* **2006,** *8,* 4053.

[62] Gordon, E.J.; Leonard, G.A.; McSweeney, S.; Zagalsky, P.F. The C1 Subunit of α-Crustacyanin: The *de novo* Phasing of the Crystal Structure of a 40 kDa Homodimeric Protein Using the Anomalous Scattering from S Atoms Combined with direct Methods. *Acta Crystallogr. D* **2001,** *57,* 1230–1237.

[63] Cianci, M.; Rizkallah, P.J.; Olczak, A.; Raftery, J.; Chayen, N.E.; Zagalsky, P.F.; Helliwell, J.R. Structure of Lobster Apocrustacyanin A1 Using Softer X-Rays. *Acta Crystallogr. D* **2001,** *57,* 1219–1229.

[64] Habash, J.; Helliwell, J.R.; Raftery, J.; Cianci, M.; Rizkallah, P.J.; Chayen, N.E.; Nneji, G.A.; Zagalsky, P.F. The Structure and Refinement of Apocrustacyanin C2 to 1.3 Å Resolution and the Search for Differences between this Protein and the Homologous Apoproteins A1 and C1. *Acta Crystallogr. D* **2004,** *60,* 493–498.

[65] Habash, J.; Boggon, T.J.; Raftery, J.; Chayen, N.E.; Zagalsky, P.F.; Helliwell, J.R. Apocrustacyanin C1 Crystals Grown in Space and on Earth Using Vapour-Diffusion Geometry: Protein Structure Refinements and Electron-Density Map Comparisons. *Acta Crystallogr. D* **2003,** *59,* 1117–1123.

6

The structural chemistry and structural biology of colouration in marine crustacea*

John R. Helliwell[†]

School of Chemistry, University of Manchester, Manchester, M13 9PL, UK

(Received 2 November 2009; final version received 18 December 2009)

The colouration of the lobster shell, famously known from its colour change on cooking, derives from a complicated mix of astaxanthin carotenoid molecules and several proteins in complex. Crystals of various components have been known for many years, but the breakthrough in structure determination came from the use of synchrotron radiation 'softer X-rays' of wavelength 2 Å, thereby targeting the increase in the sulphur anomalous scattering and the xenon L_I absorption edge. One structure of the gene-group of proteins was thus solved, and this could be used for molecular replacement solution of the β-crustacyanin dimer complex in spite of a very high solvent content of ~80%. The crystal structure of the α-crustacyanin complex of eight β-crustacyanins still eludes us, but electron microscopy (EM) structure determination would be a way forward. At present, the molecular tuning parameters causing the 100 nm bathochromic shift of the β-crustacyanin are at least known from our work and have already stimulated considerable further research in theoretical and carotenoid chemistry. The protonation state of two critically placed histidines is of keen interest here, as well as the proximity of the two astaxanthins bound to the β-crustacyanin protein. There are wider biological implications too, including the colours of rare lobsters, where site-specific amino acid changes could be a cause and much molecular biology for colour tuning would be possible. A colour-based heat sensor could also be designed. Nutraceuticals based on the health-giving properties of astaxanthin are also of keen interest. The mimicking of the colour change solely with carotenoids is being sought, supported by our structural crystallography results of unbound carotenoids. Public interest has been especially strong not least because 'the' question: 'Why does a lobster change colour on cooking?' is known to nearly everyone. Notably, popular science-summary articles have been published in magazines. We will soon be entering a new era of ultra-bright X-ray Free Electron Lasers (XFELs) and the possibility of single molecule protein structure determination; the α-crustacyanin complex of eight β-crustacyanins would provide a test of the new methods and cross-comparison with EM. The comparisons of amino acid sequences of other crustacea colouration proteins with that of β-crustacyanin provide 3D 'homology models' and allow species and evolution comparisons. This topical review provides a summary of recent work and likely future research goals.

Keywords: astaxanthin; bathochromic shift; caroteno-protein molecular recognition; crustacyanin

Contents

* From *Crystallography Reviews*, Vol. 16, No. 3, July–September 2010, 231–242.
[†] Email: john.helliwell@manchester.ac.uk

6.1 Introduction

Biological crystallography, spectroscopy, solution X-ray scattering, and microscopy have all been applied to study the molecular basis of the colouration in lobster shell. This topical review is based on the biological plenary lecture to the BCA Young Crystallographers by the author and which presented a review of progress concentrating on recent results but set in the context of more than 50 years work. In the lobster shell, there is a complex of proteins called α-crustacyanin, which is made of β-crustacyanins, which are made up of different combinations of apoproteins as post-translationally modified proteins from two genes (A2 and A3 as well as A1, C1 and C2) (see (*1*) and references therein). In particular, a breakthrough in the structural studies came from the determination of the crystal structures of apocrustacyanin A1 at 1.4 Å resolution (*2*) and β-crustacyanin (A1 with A3 protein subunits with two, shared, bound astaxanthins) at 3.2 Å resolution (*3*). The latter (Figure 6.1) was solved by molecular replacement using apocrustacyanin A1 as the search motif. A 'molecular movie' has been calculated by linear interpolation based on these two 'end point' A1 protein structures, i.e. unbound and bound astaxanthin complex A1 protein structures and was presented (*1*) as an electronic supplementary deposition within the International Union of Crystallography (IUCr) Journals archive based in Chester, United Kingdom. The movie highlights the structural changes forced upon the carotenoid on complexation with the protein. By contrast, the

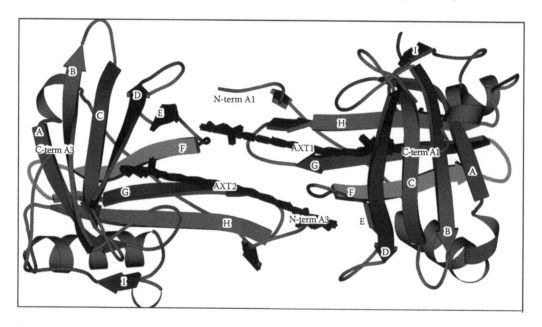

Figure 6.1 (**See colour insert.**) The structure of the A1/A3 dimer assembly (in ribbon format) with the two bound astaxanthin carotenoids (in 'stick' format). The individual β strands are shown in darker grey-black in the printed version (and in the colour insert, in the colours red, green and blue) according to the amino acid 'consensus regions' of the lipocalin protein family. Reproduced from (*3*), Copyright, 2002, National Academy of Sciences USA.

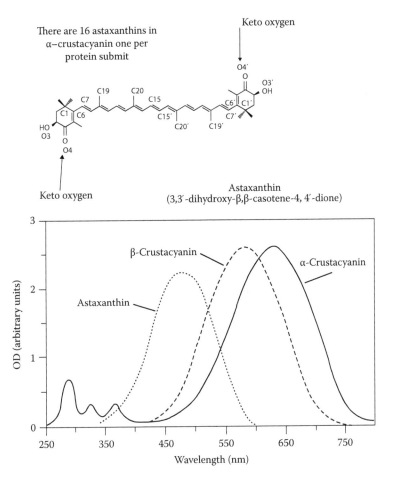

Figure 6.2 Top: The detailed chemical scheme of astaxanthin (3,3'-dihydroxy-β,β-carotene-4,4'-dione) Bottom: Absorption spectra: astaxanthin in hexane, β-crustacyanin, α-crustacyanin. Reproduced from (*1*) with permission of the International Union of Crystallography.

protein-binding site remains relatively unchanged in the binding region, but there is a large conformational change occurring in a more remote, surface loop region.

The likely molecular parameters that determine the bathochromic shift for astaxanthin (Figure 6.2) were discussed in detail (*3*). Issues remain, however, particularly associated with the electronic details including, especially, a possible protonation, or not, of two critically positioned histidine residues.

6.2 Experimental crystallography

This topical review concentrates on the crystallography and relevant experimental details are summarized here. Biochemical separations were done by Peter Zagalsky (*4*). The β-crustacyanin was then crystallized by Naomi Chayen (*1* and references therein) using a technique of crystal growth under oil known as 'microbatch' over a 4-month period, since the classic methods of crystallization repeatedly failed. The crystals were of a distinctive blue colour, like the pure protein. Cianci et al. (*3*) thereby elucidated the crystal structure of the β-crustacyanin at 3.2 Å using X-ray crystallographic techniques, including innovative use of softer X-rays of wavelength 2 Å on the synchrotron radiation source (SRS) at Daresbury Laboratory, to establish first the structure of apocrustacyanin A1 (*2*) based on instrument development work conducted some 20 years earlier (for a summary, see (*5*)).

This used a xenon heavy atom derivative prepared by placing a crystal in a high pressure of xenon gas. Four individual inert xenon gas atoms, each occupied four separate pockets within the protein. The chirality of the apocrustacyanin crystal was established using sulphur anomalous X-ray scattering also using softer X-rays. The apocrustacyanin A1 model at 1.4 Å resolution served as a molecular replacement motif to solve the β-crustacyanin structure in detail, which is comprised of an identical A1 protein subunit and a sufficiently closely related apocrustacyanin A3 protein, which has 40% amino acid sequence identity to the A1 protein.

Studying a protein at 3.2 Å is capable of yielding atomic positions and atomic displacement parameters, but it is not the easiest of territories for protein crystal structure interpretation, notably the bound water molecules were a challenge. However, the work was assisted by some technical advantages. First, the apocrustacyanin A1 model was at 1.4 Å diffraction resolution, at which level, the individual atoms are nearly separately resolved (which actually occurs at 1.2 Å or better diffraction resolution). Second, the β-crustacyanin, being composed of the two related A1 and A3 proteins, allowed structural chemistry cross-checks to be made. Namely the amino acid sequence identity in the astaxanthin-binding regions was better than the 40% value overall, referred to above, so that common structural details were available for examination. Third, for the astaxanthin itself, there was a structure derived from a true atomic resolution (i.e. standard small molecule crystallography quality) canthaxanthin structure (6). Thus, the changes to this unbound astaxanthin conformation could be elucidated rather than having to start from no structural information at all. Fourth, a particular interest turned out to be the discovery of a bound water molecule at one end of each astaxanthin at equivalent locations and at the other end a histidine for each astaxanthin (Figure 6.3). The evidence for bound waters at such a diffraction resolution is the most severe test of a 3.2 Å diffraction analysis. Evaluation was made by comparing two types of difference electron density map, namely, $2F_{obs} - F_{calc}$ and $F_{obs} - F_{calc}$, which is 'standard good protein crystallographic practice'. In addition, the correspondence of equivalent chemical environments of each of these waters was encouraging. Finally, the cross-checking for identically positioned bound waters in the superior diffraction, 1.4 Å, resolution apocrustacyanin A1 structure that had been determined, which were indeed present within 1 Å, was then overall compelling evidence. The protonation state of the histidines also was of especial interest but very difficult to resolve with X-rays, way beyond the resolution of this X-ray analysis. A more favoured approach, neutron Laue macromolecular crystallography, to pin down hydrogen details (as deuteriums), however, requires bigger crystals than were available, and fully deuterated protein. These represent current practical difficulties but obviously help to define future research avenues, including protonation prediction (7) and other biological marine systems (8).

6.3 Public interest

There has been considerable public interest in the publication (3) and the question – 'Why do lobsters change colour on cooking?' – was a stimulus to the subsequent research. The effect of cooking lobster can be reasonably guessed at – being to denature the crustacyanin unit so much that the astaxanthin becomes permanently liberated into a free (or perhaps more correctly freer) form, coloured orange at this level of dilution (noting that astaxanthin bulk solid appears as a deep purple colour). The level of public interest is reflected by the fact that the publication (3) was downloaded 634 times as a pdf in the first 6 months. As well as the TV, radio and newspaper interest (e.g. (9)), SWISSProt (10) and *Physics Today* (11) provided excellent popularization of science resumés, including a description of the molecular basis of the narrowing of the energy gap between the highest occupied and the lowest unoccupied molecular orbitals (the 'HOMO LUMO' energy levels gap).

6.4 Mimicking nature?

Recently, investigations complementary to the protein structure studies yielding several new unbound carotenoid crystal structures have allowed different candidate tuning parameters to be explored. Also, these crystal structures of free carotenoids are considerably more precise than the

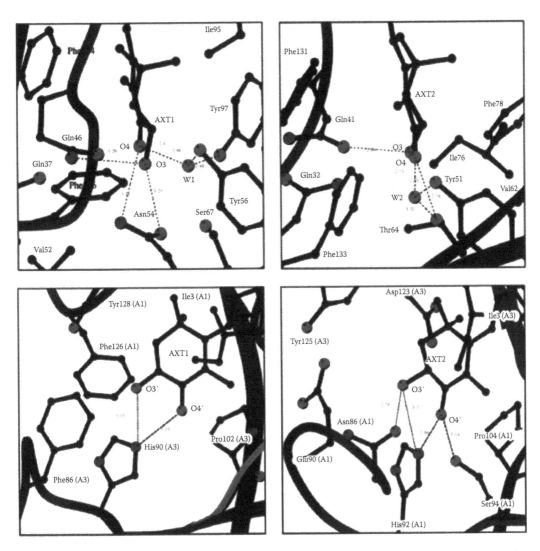

Figure 6.3 (**See colour insert.**) The detailed layout of the astaxanthin-binding sites at the end-ring molecular environments in β-crustacyanin (based on the crystal structure presented in Cianci et al. (*3*)). Top left: C1–6 end ring of AXT1 bound to the A1 molecule; top right: C1–6 end ring of AXT2 bound to the A3 molecule; bottom left: C1′–6′ end ring of AXT1 bound to the A3 molecule; bottom right: C1′–6′ end ring of AXT2 bound to the A1 molecule. Each astaxanthin molecule is equally shared between the two subunits A1 and A3. Copyright, 2002, National Academy of Sciences USA.

AXT bound to the protein. These studies include AXT in different crystal forms, as well as ester derivatives and several closely related carotenoid molecules. These crystals have a wide variety of crystal packing environments. This ensemble of crystal structures has been reported in (*12, 13*), allowing tests for the effects of carotenoid conformation, as well as the crystal packing arrangements, and to some extent, the effect of solvation, on the colours of the crystals. These results have shown that the effect of the end-ring conformation twist (Figure 6.4) on the colour of the crystals is providing a significant part of the overall bathochromic shift seen in the protein-bound AXT in β-crustacyanin, the crystals of the s-*cis* and s-*trans* conformers of astaxanthin diacetate showing a colour change (*13*) (Figure 6.5).

The bathochromic shift of approximately 50 nm, seen in the solid-state spectrum of s-*trans*-(**1**), compared with that of s-*cis*-(**1**) (Figure 6.6), can be attributed to the coplanarization of the end

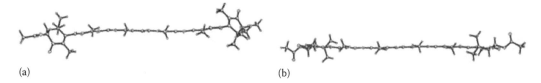

(a)   (b)

Figure 6.4 Crystal structure plots of the molecules viewed down the planes of the polyene chains (a) of the s-*cis* conformer of astaxanthin diacetate, s-*cis*-(**1**); only the higher-occupancy component is shown and solvent molecules are omitted for clarity; (b) of the s-*trans* conformer of astaxanthin diacetate, s-trans-(**1**); only the higher-occupancy component is shown. Reproduced from (*13*) with permission of the authors and the IUCr.

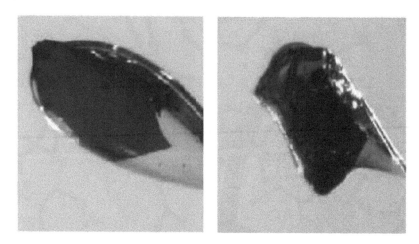

Figure 6.5 (**See colour insert.**) Crystals of the s-*cis* (left) and s-*trans* (right) conformers of astaxanthin diacetate (held in their loops for X-ray data collection). Reproduced from (*13*) with permission of the authors and the IUCr.

rings, leading to an increase in conjugation. Thus, this ensemble of crystal structures has shown here that the effect of the end-ring conformation on the colour of the crystals is providing part of the overall bathochromic shift seen in the protein-bound AXT in β-crustacyanin. This leads to the conclusion that one of the other colour-tuning mechanisms must be at work to produce the remainder of the bathochromic shift seen in α-CR and β-CR.

6.5 The theoretical chemical basis of the colour shift

These crystal structure studies have attracted considerable interest from theoretical and computational chemists. A number of different theories have been put forward as to how this very large bathochromic shift occurs (*14–19*). These include:

1. The increased conjugation arising from coplanarization of the end-rings with the polyene chain
2. An electronic polarization effect stemming from hydrogen bonding to the keto-O atoms (*14–16*)
3. Exciton coupling arising from the close proximity of the two bound astaxanthins (*18, 19*).

Most recently, Strambi and Durbeej (*20*) have analysed the exciton interaction possibilities and checked the effect in particular of the closeness of the two astaxanthins (Figure 6.7). Thus, they see in their calculation an effect of increased bathochromic shift as the two carotenoids approach each other, but rather small compared with the overall 0.5 eV effect required.

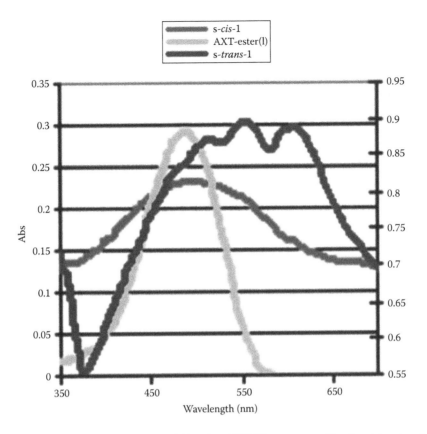

Figure 6.6 **(See colour insert.)** Solution-state (chloroform) UV-Vis spectrum of the diacetate ester of astaxanthin (green curve) and solid-state UV-Vis spectra of the s-*cis* and s-*trans* conformers, s-*cis*-**(1)** and s-*trans*-**(1)**. Reproduced from (*13*) with permission of the authors and the IUCr.

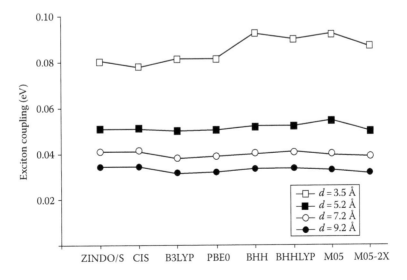

Figure 6.7 Exciton coupling in the AXT dimer at different intermolecular separations and levels of theory. Reproduced from (*20*) with permission of the authors and the American Chemical Society (ACS).

These calculations concur with the ensemble of carotenoid crystal structures (*12, 13*) where the carotenoids are still red even when at relatively close separations of around 3.5 Å.

A critical question was whether each of the two histidines at one end of each astaxanthin is protonated or not. The β-crustacyanin (*3*) X-ray crystal structure does not determine directly the protonation states of the histidine residues due to the resolution (3.2 Å) of the crystal structure. Furthermore, the protein is likely to be out of the reach of ultra-high resolution X-ray or neutron crystallography due to the large unit cell, high solvent content and small crystal size. The corresponding pK_a values have not been determined using NMR titration either. For these reasons, it is of interest to be able to predict protein protonation states of these histidine residues in particular. A double protonation prediction of these residues would further support the polarization hypothesis. Hence, Fisher et al. (*7*), i.e. Chapter 5 in this book, have undertaken predictions of these protonation states as well as an evaluation of the prediction tools themselves. Most recently time-resolved UV/Vis spectroscopy supported by theoretical calculations suggest the role for the astaxanthin as an extended enolate, deprotonated within the protein, and which would yield the colour shift mechanism.*

6.6 Conclusions and future directions

Crystals of the full complex of α-crustacyanin have been obtained but do not diffract to a high enough resolution to yield detailed structural results (*1*). Thus, electron microscopy studies firstly using uranyl acetate stain methods are a promising next step, ideally to be combined with small angle X-ray scattering methods (SAXS); SAXS data have been recorded (*1*).†

Crustacyanin is just one example of the many carotenoproteins that have been found in a variety of invertebrate animals such as asteriarubin, linckiacyanin, the carotenoprotein of the Western Rock Lobster, crustochrin, ovoverdin, ovorubin and Velellacyanin. All would be interesting to study in detail, but there are some with properties that seem to reveal features that are not consistent with the crustacyanin colouration mechanism. Detailed crystallographic (and NMR) studies such as those undertaken with crustacyanin would be interesting and informative (*21, 22*). The continued expansion of chemical crystallography carotenoid crystal structures also will extend the precise structures available and their respective crystal packing environments for a detailed dissection of structure–colouration behaviours (*23*). These will thereby offer a variety of test systems for further theoretical chemistry calculations. A fundamental question is: why lobsters have evolved such a complicated method for colouring their shells? Is it to do with avoiding a predator? An alternative explanation might be that the lobster is storing astaxanthin this way. Wade et al. (*24*) conclude 'Given the paramount importance of crustacyanin in crustacean shell colours and patterns and the critical role these play in survival, reproduction, and communication, we submit that the origin of the crustacyanin gene family early in the evolution of malacostracan crustaceans significantly contributed to the success of this group of arthropods'.

Acknowledgements

I thank SRS Daresbury Laboratory and the Joint Biology Programme of the UK Research Councils for synchrotron radiation beamtime at the STFC Daresbury Laboratory since 1981. The relevant synchrotron methods' development for softer X-rays utilization was conducted by the author whilst being a Station Scientist on SRS 7.2 between 1981 and 1984 and which was subsequently extended in the work that led to (*2*). The crustacyanin structural studies especially benefited from the award of a research grant from The Leverhulme Trust (to J.R. Helliwell, N.E. Chayen and P.F. Zagalsky), to whom I am very grateful. I thank EPSRC and the University of Manchester for a PhD studentship

---

* See Begum, S.; Cianci, M.; Durbeej, B.; Falklöf, O.; Hädener, A.; Helliwell, J.R.; Helliwell, M.; Regan, A.C.; Watt, C.I.F. On the origin and variation of colors in lobster carapace. *Phys. Chem. Chem. Phys.*, **2015**, 17, 16723–16732.

† These analyses have been undertaken and are published: Rhys, N.H.; Wang, M.-C.; Jowitt, T.A.; Helliwell, J.R.; Grossmann, J.G.; Baldock, C. Deriving the ultrastructure of α-crustacyanin using lower-resolution structural and biophysical methods. *J. Synchrotron Rad.* **2011**, 18, 79–83.

for M. Cianci, who was engaged on the crustacyanin research reviewed here. JRH thanks the EC for support under the Biocrystallogenesis Initiative (EBCI), which funded A. Olczak's post doctorate stay in Manchester. The Wellcome Trust and the BBSRC funded the Manchester Structural Chemistry Laboratory computing and graphics suite. The EC Marie Curie Training Scheme funded Ms. G. Bartalucci's PhD stay in our Laboratory and thereby her work with Dr. Madeleine Helliwell and myself. The Nuffield Foundation and the Institute Laue Langevin funded Mr. S. Fisher's Undergraduate studies Vacation Award and PhD studies, respectively, to whom I am very grateful. I am very grateful to George Britton (University of Liverpool), John Sutherland, Jonathan Clayden, Andrew Regan, Claire Baldock and Madeleine Helliwell (University of Manchester) for helpful discussions and collaborations. I thank all my co-authors for a very fruitful collaboration. I thank the Manchester Literary and Philosophical Society for permission to reproduce here parts of an earlier article published in their 'Manchester Memoirs'.

Notes on the contributor

John R. Helliwell (pictured with his bicycle before the derby match of 20th September 2009 between Manchester United and Manchester City at Manchester United's statue of George Best, Denis Law and Bobby Charlton, courtesy of Bernard McGrath) is Professor of Structural Chemistry at the University of Manchester since 1989. He has longstanding research links with Daresbury Laboratory and is currently a scientific adviser to the STFC CLIK's Knowledge Transfer to Industry. He was awarded a BA and a DSc in Physics from the University of York in 1974 and 1996, respectively, and a DPhil from the University of Oxford in 1978 and is a Fellow of the Institute of Physics since 1986, of The Royal Society of Chemistry since 1995 and of the Institute of Biology since 1998 (now the Society of Biology). He is an Honorary Member of the National Institute of Chemistry, Ljubljana, Slovenia, since 1997. He is active in serving the wider crystallographic community, e.g. through the IUCr, the European Crystallographic Association and the British Crystallographic Association, as well as SR and neutron facilities *via* their advisory committees. He is very pleased to have had the opportunity to present a Plenary Lecture to the BCA's Young Crystallographers' Group.

References

[1] Chayen, N.; Cianci, M.; Grossmann, J.G.; Habash, J.; Helliwell, J.R.; Nneji, G.A.; Raftery, J.; Rizkallah, P.J.; Zagalsky, P.F. Unravelling the Structural Chemistry of the Colouration Mechanism in Lobster Shell. *Acta Crystallogr.* **2003,** *D59,* 2072–2082.

[2] Cianci, M.; Rizkallah, P.J.; Olczak, A.; Raftery, J.; Chayen, N.E.; Zagalsky, P.F.; Helliwell, J.R. Structure of Apocrustacyanin A1 Using Softer X-rays. *Acta Crystallogr.* **2001,** *D57,* 1219–1229.

[3] Cianci, M.; Rizkallah, P.J.; Olczak, A.; Raftery, J.; Chayen, N.E.; Zagalsky, P.F.; Helliwell, J.R. The Molecular Basis of the Coloration Mechanism in Lobster Shell: β-Crustacyanin at 3.2 Å Resolution. *Proc. Natl. Acad. Sci. USA* **2002**, *99*, 9795–9800.

[4] Zagalsky, P.F. Invertebrate Carotenoproteins. *Methods Enzymol.* **1985**, *111*, 216–247.

[5] Helliwell, J.R. Overview and New Developments in Softer X-ray (2Å < λ < 5Å) Protein Crystallography. *J. Synchrotron Rad.* **2004**, *11*, 1–3.

[6] Bart, J.C.J.; MacGillavry, C.H. The Crystal and Molecular Structure of Canthaxanthin. *Acta Crystallogr.* **1968**, *B24*, 1587–1606.

[7] Fisher, S.J.; Wilkinson, J.; Henchman, R.H.; Helliwell, J.R. An Evaluation Review of the Prediction of Protonation States in Proteins Versus Crystallographic Experiment. *Crystallogr. Rev.* **2009**, *15* (4), 231–259.

[8] Zagalsky, P.F. Crustacyanin, the Blue-Purple Carotenoprotein of Lobster Carapace: Consideration of the Bathochromic Shift of the Protein-Bound Astaxanthin. *Acta Crystallogr.* **2003**, *D59*, 1529–1531.

[9] Liverpool Daily Post (Wednesday 31 July 2002). (Similar articles appeared in *The Times, The Guardian, The Independent,* on radio and TV and in *Scientific American, New Scientist* and *Physics Today*).

[10] Gerritsen, V.B. Squeeze me Swiss-Prot. *Protein Spotlight* **2002**, *26*, http://web.expasy.org/spotlight/ back_issues/026/ (accessed 2 August 2015).

[11] Day, C. Why do lobsters change colour when cooked? *Physics Today* **2002**, *November*, 22, http://scitation. aip.org/content/aip/magazine/physicstoday/article/55/11/10.1063/1.1534998 (accessed 2 August 2015).

[12] Bartalucci, G.S.; Coppin, J.; Fisher, S.J.; Hall, G.; Helliwell, J.R.; Helliwell, M.; Liaaen-Jensen, S. Unravelling the Chemical Basis of the Bathochromic Shift in the Lobster Carapace: New Crystal Structures of Unbound Astaxanthin, Canthaxanthin and Zeaxanthin. *Acta Crystallogr.* **2007**, *B63*, 328–337.

[13] Bartalucci, G.; Fisher, S.; Helliwell, J.R.; Helliwell, M.; Liaaen-Jensen, S.; Warren, J.E.; Wilkinson, J. X-ray Crystal Structures of Diacetates of 6-s-*cis* and 6-s-*trans* Astaxanthin and of 7,8-Didehydroastaxanthin and 7,8,7′,8′-Tetradehydroastaxanthin: Comparison with Free and Protein Bound Astaxanthins. *Acta Crystallogr.* **2009**, *B65*, 238–247.

[14] Weesie, R.J.; Verel, R.; Jansen, F.J.H.M.; Britton, G.; Lugtenburg, J.; de Groot, H.J.M. C-13 Magic Angle Spinning NMR Analysis and Quantum Chemical Modeling of the Bathochromic Shift of Astaxanthin in Alpha-Crustacyanin, the Blue Carotenoprotein Complex in the Carapace of the Lobster Homarus gammarus. *Pure Appl. Chem.* **1997**, *69*, 2085–2090.

[15] Durbeej, B.; Eriksson, L.A. On the Bathochromic Shift of the Absorption by Astaxanthin in Crustacyanin: A Quantum Chemical Study. *Chem. Phys. Lett.* **2003**, *375*, 30–38.

[16] Durbeej, B.; Eriksson, L.A. Conformational Dependence of the Electronic Absorption by Astaxanthin and its Implications for the Bathochromic Shift in Crustacyanin. *Phys. Chem. Chem. Phys.* **2004**, *6*, 4190–4198.

[17] Durbeej, B.; Eriksson, L.A. Protein-Bound Chromophores Astaxanthin and Phytochromobilin: Excited State Quantum Chemical Studies. *Phys. Chem. Chem. Phys.* **2006**, *8*, 4053–4071.

[18] van Wijk, A.A.C.; Spaans, A.; Uzunbajakava, N.; Otto, C.; de Groot, H.J.M.; Lugtenburg, J.; Buda, F. Spectroscopy and Quantum Chemical Modeling Reveal a Predominant Contribution of Excitonic Interactions to the Bathochromic Shift in Crustacyanin, the Blue Carotenoprotein in the Carapace of the Lobster Homarus Gammarus. *J. Am. Chem. Soc.* **2005**, *127*, 1438–1445.

[19] Ilagan, R.P.; Christensen, R.L.; Chapp, T.W.; Gibson, G.N.; Pascher, T.; Polivka, T.; Frank, H.A. Femtosecond Time-Resolved Absorption Spectroscopy of Astaxanthin in Solution and in α-Crustacyanin. *J. Phys. Chem. A* **2005**, *109*, 3120–3127.

[20] Strambi, A.; Durbeej, B. Excited-State Modeling of the Astaxanthin Dimer Predicts a Minor Contribution from Exciton Coupling to the Bathochromic Shift in Crustacyanin. *J. Phys. Chem. B* **2009**, *113*, 5311–5317.

[21] Helliwell, J.R.; Helliwell, M. Unravelling the Chemical Basis of the Bathochromic Shift of the Lobster Carapace Carotenoprotein Crustacyanin. In *Models, Mysteries and Magic of Molecules:* Boeyens, J.C.A., Ogilvie, J.F., Eds.; Springer: New York, 2008; pp. 193–208.

[22] Britton, G.; Helliwell, J.R. Carotenoid-Protein Interactions. In *Carotenoids: Natural Functions:* Britton, G., Liaaen-Jensen, S., Pfander, H., Eds.; Birkhauser Verlag: Basel-Boston-Berlin, 2008; Vol. 4, pp. 99–118.

[23] Helliwell, M. Three-dimensional Structures of Carotenoids by X-ray Crystallography. In *Carotenoids: Natural Functions:* Britton, G., Liaaen-Jensen, S., Pfander, H., Eds.; Birkhauser Verlag: Basel, 2008; Vol. 4, pp. 37–52.

[24] Wade, N.M.; Tollenaere, A.; Hall, M.R.; Degnan, B.M. Evolution of a Novel Carotenoid–Binding Protein Responsible for Crustacean Shell Color. *Mol. Biol. Evol.* **2009**, *26*(8), 1851–1864.

7 Where is crystal structure analysis heading in the future?

If there is a future International Year of Crystallography, and 'why not?' since our turn will surely come round again, let's say at the bicentennial in 2112, what might we see then?

The databases are a reflection of the state of crystal structure analysis and a health metric is the time period over which a doubling of their entries occurs. If we assume a doubling period of 10 years, then in 100 years from now, a theoretical 2^{10} multiplication of entries would promise, if realized, an obviously amazing increase. In terms of the methods of macromolecular de novo crystal structure determination, the last 20 years researchers have veered away from multiple isomorphous replacements through multi-wavelength anomalous dispersion methods of phase determination to, today, molecular replacement being predominant. This trend seems set to continue. I continue to believe however in the vital importance of the use of resonant scattering in phase determination for macromolecules and for precise element identification in chemical crystallography (especially mixed metal sites) and in biological crystallography [1–3] whether using synchrotron radiation or in the newcomer on the block, the X-ray laser [4].

In terms of database contents, given the myriad number of proteins, their complexes and their ligands as well as their permutations and combinations that are possible, there apparently will be no limit to the number of new crystal structures. There will only be a limit set by our ingenuity at automation of all the steps of a crystal structure determination. Finally, the number of time-resolved crystallography studies is increasing, and albeit special cases, nevertheless, give a remarkable insight from experiment into the molecular dynamics stimulated by light or other reaction initiators in the crystal. There is a synergy of such experimental and computational molecular dynamics. Indeed as computational simulations stretch now to microsecond time scales, the link of these to help explain the diffuse scattering from a crystal should revive, and hopefully in earnest.

The limitations of needing a crystal are also changing. For the large biological complexes, single particle cryo-electron microscopy is seeing a spurt of development though improved sensitivity electron detectors and the jitter-blurring in the EM images being correctable by adjusting the spatial alignment of recorded timeframes. The X-ray laser has moved through increasingly smaller and smaller samples towards the vision of single molecule diffraction, and which is still a declared objective of those facilities.

In the objective of realizing a biological structure, complete with its hydrogen atoms via neutron protein crystallography, the traditional challenge has been the need for relatively big protein crystals. This need has steadily reduced as initiatives in the methods such as using the white beam of emitted neutrons have helped as well as microbiological protein expression on fully deuterated media have basically doubled the number of significantly scattering atoms (i.e. the effect that deuterium scatters as well as carbon). Assessments of the possible changes to protein structure and/or the ionization properties of deuteration have been examined in a wide range of cases and found hardly ever to occur. (The kinetics of deuterated chemistries is affected obviously as deuterium is heavier than hydrogen.)

Crystallisation methods, which obviously underpin all that we do as crystallographers, have steadily improved too in the last decades. Thus, whilst in the past, protein crystal growth was typically known as a 'black art', it has become a scientific and technologically adept area of science.

Closely related to crystallography are the complementary methods involving scattering from solutions of macromolecules: Small Angle X-ray and Small Angle Neutron Scattering (SAXS and SANS) methods. Thus, overall details such as the radius and shape of a macromolecule can be determined. Time-resolved studies are most readily done in the solution state, i.e. where large conformational changes are possible. The neutrons case, SANS, can also harness so-called contrast variation by varying the percentage of heavy and light water or by partial deuteration of a complex of protein and nucleic acid.

A further method is used to study the fine details of metalloproteins, namely, X-ray absorption spectroscopy (XAS), i.e. with the X-ray wavelength tuned to the X-ray absorption edge of the inherent metal atom. The XAS signal from the metal is exquisitely sensitive to the immediate neighbouring ligand atoms to the metal atom and also requires relatively small X-ray doses to measure the X-ray spectrum. A modern case is the study by XAS and by X-ray crystallography of the photosynthesis protein 'PSII' and its cluster of manganese atoms (the 'oxygen-evolving complex') at the heart of its catalytic splitting of water; these two techniques have yielded complementary information.

In a look-back of the last 100 years, it is widely stated that the X-ray fibre diffraction studies of DNA and the subsequent molecular model for the double helix and its explanation of the basis of heredity was the most exciting breakthrough of twentieth century science. Therefore, that scientific discovery and its discoverers (Watson and Crick) surely rival the stature of Isaac Newton and the theory of gravitation. In 2112, what might we as crystallographers have to offer the competition of 'best discovery' of the next 100 years? That choice I leave to the reader of this book to contemplate, but I think it safe to assume that we crystallographers will be taking part in that competition. Crystallography has a very bright future.

REFERENCES

1. Cianci, M., Helliwell, J. R., Helliwell, M., Kaucic, V., Logar, N. Z., Mali, G. and Tusar, N.N. Anomalous scattering in structural chemistry and biology. *Crystallogr. Rev.* **11**, 245–335, 2005.
2. Helliwell, J.R., Helliwell, M., Kaucic, V. and Logar, N.Z. Resonant elastic X-ray scattering in chemistry and materials science. *Eur. Phys. J.* (Special Topics) **208**, 245–257, 2012.
3. Helliwell, J. R. Resonant elastic X-ray scattering in life science, chemistry and materials science; recent developments. REXS 2013—Workshop on Resonant Elastic X-ray Scattering in Condensed Matter. IOP Publishing. *J. Phys. Conf. Ser.* **519**, 2014. 012002 doi:10.1088/1742-6596/519/1/012002.
4. Helliwell, J. R. Viewpoint review of de novo protein crystal structure determination from X-ray free-electron laser data. *Crystallogr. Rev.* **20**(3), 207–209, 2014; Barends, T. R. M., Foucar, L., Botha, S., Doak, R. B., Shoeman, R. L., Nass, K., Koglin, J.E., Williams, G. J., Boutet, S., Messerschmidt, M. and Schlichting, I. De novo protein crystal structure determination from X-ray free-electron laser data. *Nature* **505**, 244–247, 09 January 2014. doi:10.1038/nature12773.

Section IV

Societal impacts

8 Crystallography and sustainability

8.1 INTRODUCTION

First, what is 'sustainability'? Wikipedia provides a readily available and helpful description, see http://en.wikipedia.org/wiki/Sustainability:

'In ecology, *sustainability* is how biological systems remain diverse and productive.'

Wikipedia provides further details:

'The world's sustainable development goals are integrated into the eight *Millennium Development Goals (MDGs)* that were established in 2000 following the Millennium Summit of the United Nations. Adopted by the 189 United Nations member states at the time and more than twenty international organizations, these goals were advanced to help achieve the following sustainable development standards by 2015:

1. To eradicate extreme poverty and hunger
2. To achieve universal primary education
3. To promote gender equality and empower women
4. To reduce child mortality
5. To improve maternal health
6. To combat HIV/AIDS, malaria and other diseases
7. To ensure environmental sustainability
8. To develop a global partnership for development

According to the data that member countries represented to the United Nations, Cuba was the only nation in the world in 2006 that met the World Wide Fund for Nature's definition of sustainable development, with an ecological footprint of less than 1.8 ha per capita, 1.5, and a Human Development Index of over 0.8, 0.855.'[*,†]

Crystallography most obviously is assisting with goal 6. Examples and case studies are cited in Appendix 8.B of this chapter.

Furthermore, crystallography has, since its earliest period when William Henry Bragg had a substantial fraction of his research students being female, had an excellent proportion of its researchers being female relative to other areas of science. These included Kathleen Lonsdale, Dorothy Hodgkin and Sine Larsen as female Presidents of the International Union of Crystallography (1966, 1981–1984 and 2008–2011 respectively), and up to Nobel Prize winner level with Dorothy Hodgkin (Nobel Prize in Chemistry 1964) and Ada Yonath (Nobel Prize in Chemistry 2009). Crystallography research therefore has assisted in goal 3. I have endeavoured to assist in this process via my work as Gender Equality Champion at the School of Chemistry in the University of Manchester via the Athena SWAN framework of the UK Government's Equality Challenge Unit (see http://www.chemistry.manchester.ac.uk/media/eps/schoolofchemistry/aboutus/athenaswan/pdf/ManchesterChemistry_AthenaSWAN_Application_BronzeAward.pdf, accessed 2 August 2015).

---

[*] "Living Planet Report 2006". World Wide Fund for Nature, Zoological Society of London, Global Footprint Network. 24 October 2006. p. 19. Retrieved 18 August 2012.

[†] World failing on sustainable development.

Goal 2: Assisting the realization of the goal *To achieve universal primary education*, is less easy perhaps as crystallography is in its details built around the core sciences and mathematics. However, as I have described in Chapter 1, this is not an impediment in finding ways of reaching out to the Public and to school children. 'Primary education' however presumably means 'reading, writing and arithmetic' basic skills. When I started my enquiries about giving a lecture on crystallography in prisons, I was alerted to the quite often low education skills I might expect in my audience. As I have yet not managed to give a lecture to this audience, my practical knowledge of how to proceed is at the embryonic stage. My ideas include to especially take advantage of the highly visual materials that one has available especially crystals (calcite, quartz and so on) and molecular models.

8.2   RESEARCH EXAMPLES RELEVANT TO SOCIETY

In the lecture that I prepared with Brian McMahon of the International Union of Crystallography in Chester, United Kingdom, for the New Delhi CODATA 2014 Congress and General Assembly (see Figure 8.1 for the conference banner headline), we summarised the topic:

Sustainability of life and molecular crystallography 3D data

We focussed on crystallography's role in improving health and efficient energy usage which are and can be increasingly at the molecular level based on crystal structure analyses. In health sciences, structure-based drug design is routinely employed; in energy research, hydrogen storage is addressed also using 3D atomic structures. Thus, crystallography has an important role within the topic of the sustainability of Life.

A key part of this process of broad applicability rests on world wide access to the three-dimensional (3D) structures database of biological macromolecules (Protein Data Bank) and the chemical crystallography databases (CCDC, ICDD, COD, Metals). These provide fantastic resources spanning a vast range of the living and material world in atomic detail. The tools of physics (X-rays, neutrons and electrons in diffraction, microscopies and spectroscopies) allow us to determine and study in detail these structures and their atomic interactions. Clear technical understanding of these physical methods provides clarity on how much trust we can place in the correctness of these structures, i.e. their precision and accuracy.

Furthermore, life under 'normal' and extreme conditions can be reliably compared: hot springs, high saline and extreme cold examples of protein crystal structures have been determined. It is a natural next step to understand the effects of pollution; a simple example is the binding of nitrates that would come from excessive use of such fertilisers (Figure 8.2).

Crystallography has a long tradition in community methods development, and in sharing derived and processed data at the databases referred to above. Research journals in this field have long provided links to derived datasets (molecular structures), and in more recent years to processed experimental data (single crystal X-ray structure factors and powder diffraction profiles), and in turn, there have been links from curated datasets in structural databases back to the associated publication [1].

The journals have *stringent requirements* that articles will only be published with the accompanying data and/or, in the life sciences, Protein Data Bank deposition files (derived coordinates and processed structure factor amplitudes). At the submission stage of an article, information should be provided to convince the referees that the interpretations of the diffraction data and electron-density

SciDataCon 2014
International Conference on
Data Sharing and Integration for Global Sustainability

FIGURE 8.1   The conference banner for the CODATA conference 2014.

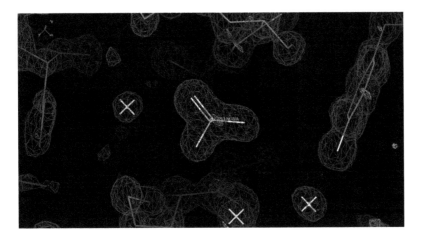

FIGURE 8.2 **(See colour insert.)** Discrete nitrate ion binding to a protein surface. In this molecular graphics representation, the blue mesh is the electron density derived from the X-ray diffraction from a single crystal of a protein (lysozyme in a triclinic crystal form). The interpretation of this in terms of chemistry then can readily be made: on the left, a nitrate ion and on the right, a tyrosine amino acid.

maps and/or structures are correct, within the resolution of the analysis; in the life sciences, a Protein Data Bank structure validation report is also required on submission. Dataset requirement *recommendations* exist for articles that present experimental small-angle X-ray or neutron scattering data. For these, the deposition of an ASCII file representing the background-corrected scattering profile(s) with errors is recommended. For powder diffraction, articles that present the results of powder diffraction profile fitting or refinement (Rietveld) methods require deposition of the primary diffraction data, i.e. the numerical intensity of each measured point on the profile as a function of scattering angle. For the chemical sciences (which obviously includes many small-molecule ligands of biological macromolecules), the requirements on derived coordinates and processed diffraction data are more stringent, and such data are held for example within the IUCr publication archive. For example, among the IUCr Journals that publish such data, this quote (from the Notes for Authors of *Acta Crystallographica Section E: Structure Reports Online*) is typical: 'Supporting information (such as experimental data, additional figures and multimedia content) that may be of use or interest to some readers but does not form part of the article itself will be made available from the IUCr archive. Arrangements have also been made for such information to be deposited, where appropriate, with other relevant databases' [2]. The latest discussions within IUCr are now looking towards archiving of raw data [3] and one of the first examples of such is described in Ref. [4]. *The Australian synchrotron data archive 'Store.Synchrotron' is an exemplar, whose work includes its support of users with publications having raw datasets, and its ability to manage digital object identifier (DOI) registration. This archive also releases raw diffraction image datasets for public analysis* [5].

The above description is fairly detailed, but it shows the rigorous efforts made by crystallographers to archive their results along with the metadata being provided as well to describe their data thoroughly. Preservation of data ensures the sustainability of the crystallographic research effort (along with the crystallographic science publications). This is proving an important role within extensive efforts to describe nanomaterials properly, essential for their rigorous safety description. There is a CODATA/VAMAS Joint Working Group on the Description of Nanomaterials (http://www.codata.org/nanomaterials) and from which I quote:

> Nanomaterials are complex, and researchers continue to develop new and innovative materials. Describing nanomaterials is a challenge for all user communities, but a description system is essential to ensure that everyone knows exactly which nanomaterial is being discussed, whether for research, regulatory, commercial, or other purposes.

CODATA and VAMAS, an international pre-standardization organization concerned with materials test methods, have set up a joint working group to help develop a uniform description system for nano-materials. This international working group includes representatives from virtually every scientific and technical discipline involved in the development and use of nanomaterials, including physics, chemistry, materials science, pharmacology, toxicology, medicine, ecology, environmental science, nutrition, food science, crystallography, engineering, and more. Many international scientific unions actively participate. The IUCr is an active participant. Nanomaterials are one of the most actively pursued areas of modern day materials research.

A major feature of crystallography in the life sciences is the use of synchrotron radiation in macromolecular structure determination (see Appendix 8.B with impact case studies and publication examples).

8.3   CRYSTALLOGRAPHERS AND PEACE

It is striking to me that crystallographers have taken a prominent role trying to ensure Peace. There is surely a crucial role for sustainability of preserving peace; there surely should be a bullet point for World Peace in the United Nations' list quoted at the start of this Chapter!

Kathleen Lonsdale (President of the IUCr 1966) wrote a slim but powerfully argued book entitled *Is Peace Possible?* https://archive.org/details/ispeacepossible00lons. She went to jail during the Second World War for her beliefs as a pacifist.

Dorothy Hodgkin was President of Pugwash from 1976 to 1988. The Pugwash Conferences on Science and World Affairs is an international organization that brings together scholars and public figures to work towards reducing the danger of armed conflict and to seek solutions to global security threats. See http://pugwash.org/. In their *Dialogue Across Divides* they state:

> We have to learn to think in a new way. Pugwash seeks a world free of nuclear weapons and other weapons of mass destruction. We create opportunities for dialogue on the steps needed to achieve that end, focusing on areas where nuclear risks are present. Moving beyond rhetoric, we foster creative discussions on ways to increase the security of all sides in the affected regions. Remember your humanity, and forget the rest.

Linus Pauling won two Nobel Prizes, one for chemistry (1954) and one for Peace (1962). Linus Pauling was a great early practitioner of X-ray crystal structure analysis with a wide variety of his crystal structures featuring in his famous textbook on chemistry *The Nature of the Chemical Bond*. He also wrote *No More War* (1958) and specifically his concerns about the safety level limits set for radiation:

> I believe that the nations of the world that are carrying out the tests of nuclear weapons are sacrificing the lives of hundreds of thousands of people now living and of hundreds of thousands of unborn children, and that this sacrifice is unnecessary.

In the 1980s, I was at a dinner for early career scientists hosted by the then Secretary of State for Education, Kenneth Baker, and I sat next to his Principal Private Secretary Sir David Hancock (1934–2013; see e.g. http://www.theguardian.com/politics/2013/sep/29/sir-david-hancock). I explored the argument on Sir David of removing the Defence budget in favour of spending it instead on more scientific research. He replied to the effect that any such move, whilst impossible for any modern Government to contemplate, would not lead to more spending on scientific research but would be used instead to reduce taxes; sobering thoughts.

8.A   APPENDIX: EXAMPLES OF THE SOCIAL AND ECONOMIC IMPACT OF CRYSTALLOGRAPHY: TOWARDS A MORE SUSTAINABLE FUTURE FOR LIFE ON EARTH

RCUK Analysis of the impact of SRS Protein Crystallography in Knowledge Transfer: Protein Crystallography at the Synchrotron Radiation Source (SRS), Daresbury Laboratory (1981–2008; Figures 8.3 and 8.4).

FIGURE 8.3 **(See colour insert.)** The UK's Science and Technology Facilities Council (STFC) undertook an analysis of the social and economic impact of the Daresbury Laboratory Synchrotron Radiation Source, which operated from 1981 to 2008; this figure shows the front cover of the report. The full report is available at www.stfc.ac.uk or with google use the search term "SRS social and economic impact 1981–2008" (accessed 2 August 2015).

8.B   APPENDIX: REFERENCES AND THEIR BRIEF DESCRIPTION

Light Harvesting Protein (LH2) (1995): Important for understanding how sunlight is trapped by bacteria McDermott, G., Prince, S. M., Freer, A. A., Hawthornthwaite-Lawless, A. M., Papiz, M. Z., Cogdell, R. J. and Isaacs, N. W. (1995) Crystal structure of an integral membrane light-harvesting complex from photosynthetic bacteria. *Nature* 374, 517–521.

The 30S Ribosome subunit (1999–2000) resolved to 3.0 Å: Jointly with SRS and ESRF. This structure was vital to understanding how genetic information is translated into the protein amino acid sequence. Wimberly, B. T., Brodersen, D. E., Clemons, W. M. Jr., Morgan-Warren, R. J.,

FIGURE 8.4 The Research Council's UK undertook an analysis of the economic impacts of its individual component research councils and who provided exemplar cases; in the case of STFC two were chosen and one of these was 'Protein crystallography at the SRS' and from within which the highlighted publications quoted in Appendix 8.B are listed. This figure shows the front cover of the report; the full report is available at www.rcuk.ac.uk section 'Publications' access "SRS impact" (accessed 2 August 2015).

Carter, A. P., Vonrhein, C., Hartsch, T. and Ramakrishnan, V. (2000) Structure of the 30S ribosomal subunit. *Nature* 407, 327–339.

Anthrax lethal factor (2001): This crystal structure revealed how the protein inhibits one or more signalling pathways critical to the pathogenesis of the anthrax bacteria. Pannifer, A. D., Wong, T. Y., Schwarzenbacher, R., Renatus, M., Petosa, C., Bienkowska, J., Lacy, D. B., Collier, R. J., Park, S., Leppla, S. H., Hanna, P. and Liddington, R. C. (2001) Crystal structure of the anthrax lethal factor. *Nature* 414, 229–233.

Lobster β-crustacyanin at 3.2 Å resolution (2002): This crystal structure revealed the molecular basis for the colouration mechanism in lobster shell. There was large-scale media coverage of this work in both the United Kingdom and United States showing the Public's keen interest in science matters. This research has a very broad relevance to coloration of marine crustacea and probably explains the origin of one of the mechanisms of camouflage, a fundamental aspect of Life on Earth. Cianci, M., Rizkallah, P. J., Olczak, A., Raftery, J., Chayen, N. E., Zagalsky, P. F. and Helliwell, J. R. (2002) The molecular basis of the coloration mechanism in lobster shell: β-crustacyanin at 3.2 Å resolution. *Proceedings of the National Academy of Sciences USA* 99, 9795–9800.

Light-Harvesting Protein (LH1) reaction centre core complex (2003) at 4.8 Å resolution: This crystal structure contributed further to the understanding of how bacteria utilise sunlight and is one of several major steps towards arriving at alternative, more natural and friendly, energy sources in the future. Roszak, A. W., Howard, T. D., Southall, J., Gardiner, A. T., Law, C. J., Isaacs, N. W. and Cogdell, R. J. (2003) Crystal structure of the RC-LH1 core complex from Rhodopseudomonas palustris. *Science* 302, 1969–1972.

Understanding the molecular basis of the 1918 'Spanish' influenza pandemic virus haemagglutinin (2004): This crystal structure contributed to the understanding of the extraordinarily high infectivity and mortality rates observed during the 1918 flu epidemic and suggested new drug designs. Gamblin, S. J., Haire, L. F., Russell, R. J., Stevens, D. J., Xiao, B., Ha, Y., Vasisht, N., Steinhauer, D. A., Daniels, R. S., Elliot, A., Wiley, D. C. and Skehel, J. J. (2004) The structure and receptor binding properties of the 1918 influenza haemagglutinin. *Science* 303, 1838–1842.

In research led by Professor David Stuart and coworkers new ways to tackle Foot and Mouth Disease Virus (FMDV) are approaching steadily and has extensively relied on using synchrotron radiation crystallography, to begin with at the SRS and now much more rapidly and effectively at the UK's Diamond Light Source, with its considerable technical superiority of its third generation SR source design. https://www.stfc.ac.uk/files/foot-and-mouth-case-study/ (accessed 2 August 2015).

A very recent, wide ranging, discussion of crystallography and sustainability, including particular reference to green chemistry, is the American Crystallographic Associations (ACA) Transactions Symposium 2015 (6).

REFERENCES

1. McMahon, B. An integrated web resource for crystallography. *Data Science J.* **1**, 54–67, 2006.
2. Notes for authors. *Acta Crystallogr.* E**70**, 3–8, 2014, http://journals.iucr.org/e/issues/2014/07/00/me0521/index.html.
3. Helliwell, J. R. *et al.* Triennial report to the IUCr Executive Committee from the Diffraction Data Deposition Working Group (DDDWG), http://forums.iucr.org/viewtopic.php?f=21&t=343 (accessed on: 13 September 2015).
4. Tanley, S. W. M., Schreuers, A. M. M., Helliwell, J. R., Kroon-Batenburg, L. M. J. Experience with exchange and archiving of raw data: comparison of data from two diffractometers and four software packages on a series of lysozyme crystals. *J. Appl. Crystallogr.* **46**, 108–119, 2013.
5. Meyer, G. R., Aragão, D., Mudie, N. J., Caradoc-Davies, T. T., McGowan, S., Bertling, P. J., Groenewegen, D., Quenette, S. M., Bond, C. S., Buckle, A. M. and Androulakis, S. Operation of the Australian Store. Synchrotron for macromolecular crystallography. *Acta Crystallogr.* **D70**, 2510–2519, 2014.
6. American Crystallographic Associations (ACA). Crystallography and sustainability. *Transactions Symposium 2015*, edited by Lind-Kovacs, C. and Rogers, R., in press, http://www.amercrystalassn.org/content/pages/main-transactions.

Author Index

Subject Index

Milton Keynes UK
Ingram Content Group UK Ltd.
UKHW050451071024
449327UK00015B/330

9 780367 377229